Science for the Fun of It

Science for the Fun of It

A Guide to Informal Science Education

edited by Marvin Druger

National Science Teachers Association
1742 Connecticut Avenue, NW
Washington, DC 20009

This book has been produced by Special Publications
National Science Teachers Association
1742 Connecticut Avenue, NW
Washington, DC 20009

Edited by Cara Young and Crystal Hamann, associate editors
Shirley Watt, managing editor
Designed by Chroma Design and Communications,
Takoma Park, MD 20912

ISBN Number: 0–87355–074–9

Contents

Preface

School provides only a small part of our total education. We learn from everything we do, in and out of school, and everything we do becomes part of what we are. *Science for the Fun of It: A Guide to Informal Science Education* focuses on science learning outside the classroom. The book consists of a collection of articles by people who are involved intimately with informal science education. The intent is to highlight some of the many out-of-school opportunities for learning science so that readers can become more aware of these resources and use them more effectively. Such resources include zoos, museums, television, magazines and books, and a variety of creative programs and projects. We could include only a small sample of existing resources, but enough variety is here to indicate the educational value and diverse nature of these opportunities.

The educational value of informal science education extends beyond mere acquisition of knowledge. Attitudes toward science can be greatly influenced by science experiences beyond the classroom. The Natural History Museum in New York City was a favorite place for me to visit as a youngster. I felt comfortable among the dinosaurs and large fishes and mammals. I could not hope to learn everything—there was too much to observe. The museum was a warehouse full of natural history treasures that could not be memorized, but could be appreciated and enjoyed. The museum was a place where I could live and breathe the past and present of the natural world and contemplate its future. The museum imprinted me with feelings more than with information. These feelings would later spill over into formal science education in the classroom. Similarly, the resources described in this book do not replace formal science education; they supplement and enhance it.

Informal science education resources also can provide a strong foundation for learning science. Like many of you, I have always enjoyed visiting zoos. As a youngster, I didn't visit zoos to learn about animals. I went simply to see animals and to have fun, but I learned about animals in spite of my nonacademic motives. I have visited many zoos around the world. I still visit zoos to see animals and to have fun, but I learn more because I have a history of visiting zoos and a frame of reference about animals and biology. Academic learning and zoo visits have become integral parts of my science education. Thus, what we experience outside the classroom contributes in a meaningful way to our overall knowledge and appreciation of science.

Informal science education resources also appeal to peoples' imagination and curiosity. The Baltimore Aquarium is indeed a wonder. How can anyone fail to be amazed by the diversity and adaptedness of life in the sea while exploring this aquarium? I press my nose against the glass and become part of the world of undersea life. I fantasize about being attacked by a giant shark, and repulsing the attack with a punch on the nose with my bare fist. I imagine wandering the oceans with the giant sea turtles. I even see myself teaching biology to a school of brightly colored fish and giving them exams. Anything is possible as I float through the aquarium sea, stimulated by the unreality of reality.

Participating in informal science education may be pleasurable and exciting, but it is also important to think about how such resources can be used to gain the most educational benefits. I recall observing an elementary school teacher with her class of wide-eyed youngsters standing under the towering Brontosaurus reconstruction at a museum. This huge dinosaur had a very small brain. The teacher was intent upon making this point clear to her students. "Is the dinosaur's brain large or small?," she asked. In unison, the awed children responded, "It's laaarge!" "That's right," echoed the teacher, "It's smaaall!"

Such misuses of informal science teaching resources probably occur more often than one would like to see. To help make informal science education resources more useful, we have included "Helpful Hints" at the end of chapters, where appropriate. We hope that readers will refer to these suggestions to help enhance informal learning.

Science for the Fun of It is divided into four sections. The first section provides reasons for and background information on learning science outside the classroom. The second section looks at the media as a resource and includes chapters on television and science publications. The third section focuses on museums and zoos. The last section presents a variety of special projects and creative approaches to science learning outside the classroom.

Publishing this book would not have been possible without the competent assistance of the NSTA staff. In particular, I thank Shirley Watt, managing editor of NSTA's Special Publications, and Cara Young and Crystal Hamann, associate editors. George Tressel and Michael Templeton of the National Science Foundation were particularly helpful in providing suggestions. George Tressel also compiled the bibliography on informal science education. I thank Leroy Lee, past-president of NSTA, for supporting the initiation of the project. Finally, my wife Pat deserves special thanks for providing the personal and professional advice that only a spouse can provide, and because she rarely receives recognition for all the exceptional things she does.

I hope you enjoy reading this book—that it will help you become more aware of the value of science beyond school, and that such awareness and subsequent involvement will enrich your life and the lives of your students or children.

Marvin Druger, editor

Authors

Paula Apsell has been executive producer of *NOVA* since 1984. During her tenure, the series has won several major awards, including one AAAS/Westinghouse Science Journalism Award and two national Emmys. From 1980–1984 she was the senior medical producer for WCVB–TV in Boston, and she received an Ohio State Award for producing the documentary *Someone I Once Knew,* on Alzheimer's disease. From 1983–1984 Apsell studied at MIT on a Vannevar Bush Fellowship. Apsell joined the staff of NOVA in 1975, advancing from production assistant to staff producer, in which capacity she produced eight *NOVA* films.

Annette Berkovits, born in the Khirgiz Republic, U.S.S.R., received her primary education in Poland. She completed her science education in the United States, first at the Bronx High School of Science, and then at the City College of the City University of New York. She received a masters degree in educational administration and supervision from Manhattan College in New York.

Berkovits joined the staff of the New York Zoological Society in 1972. As director of education she is responsible for all of the Bronx Zoo's formal and informal interpretive services. These include adult education and school programs, curriculum development projects, audiovisual services, consultation on development of new exhibit graphics, and international conservation education. She is also responsible for managing the Camel Rides operation and the world-famous Children's Zoo. Since 1981, Berkovits has been project director/principal investigator on several major National Science Foundation grants, including the nationally and internationally recognized Project W.I.Z.E. (Wildlife Inquiry Through Zoo Education). She serves as a panelist, grant reviewer, and consultant for agencies such as The New York State Council on the Arts, the Institute of Museum Services, the National Science Foundation, and several museums. An avid world traveler and student of museum education, she has visited zoos in 16 countries. Annette Berkovits is listed in the 1988 edition of *Who's Who in American Women.*

Joel N. Bloom is president and director of the Franklin Institute Science Museum in Philadelphia. Active in both national and international museum professional activities, he is currently president of the American Association of Museums and past president of the Association of Science–Technology Centers. He chairs the U.S. National Committee of the International Council of Museums. He is past president of the Greater Philadelphia Cultural Alliance and a member of the Mayor's Cultural Advisory Council of Philadelphia. Under his leadership, the Franklin Institute has been a pioneer in training science teachers and developing and distributing hands-on science kits for classroom use. Bloom was cochair of the American Association of Museums' Commission on Museums for a New Century and a member of the Museum Advisory Panel of the National Endowment for the Arts and the National Science Foundation Advisory Committee for Science Education.

Ray Bradbury has published 23 books of stories, novels, poems, and plays since he began publishing at age 20. He began writing when he was 12 years old. Perhaps his best-known book is *The Martian Chronicles,* dealing with the colonization of Mars. In 1964 he created the historical scenario for the U.S. Pavilion at the New York World's Fair. He was creative consultant on *Spaceship Earth,* a history of communication from ancient to modern times, at Florida's EPCOT Center. He wrote the dramatic history of astronomy for the Air and Space Museum in Los Angeles, and he created the oral text for an outline of the Space Age for the U.S. Pavilion in Toronto, Ontario, in 1986. In 1954 he wrote the screenplay for John Huston's motion picture *Moby Dick.*

Michelene T. H. Chi is an associate professor of psychology and a senior scientist at the Learning Research and Development Center at the University of Pittsburgh, Pennsylvania. She received both a Spencer Fellowship and the Boyd R. McCandless Young Scientist Award for distinguished theoretical contribution and programmatic research efforts in the area of the psychology of learning. Her current research most relevant to informal science learning examines the question of how students' self-generated explanations of examples or visual displays can foster greater understanding of domain theories, as well as revise their prior naive conceptions. She has served as secretary of Division C of the American Educational Research Association, and she currently serves on the editorial boards of *Cognitive Development* and *Human Development.* Her publications include a volume she edited entitled *The Nature of Expertise.*

Valerie Crane is president of Research Communications, a company established to serve the communications needs of broadcasting, business, industry, and government. As president, Dr.

Crane brings extensive experience with a wide variety of media production agencies in both television and radio.

Dr. Crane has written a book chapter on content development in television, published by Academic Press, and has recently completed a book chapter on formative research for teletext. She has organized and conducted a series of seminars on audience research for a variety of academic institutions and production agencies and is listed in *Who's Who in American Women, Who's Who in the East,* and the *International Register of Profiles.* She also serves as a member of AAAS's Committee on Public Understanding of Science and Technology, and in 1988 she will begin her term as representative-at-large for AECT. She was also field editor of the research column of the National Association of Educational Broadcasters from 1979–1981.

In 1975, Dr. Crane became a senior research consultant at Heuristics, Inc., a research and evaluation firm based in Wellesley, Massachusetts. In 1978, she established and directed the Media Division Public Affairs Research Institute. As director of the Media Division, she obtained and managed contracts with organizations, including the Agency for Instructional Television, Corporation for Public Broadcasting, WGBH–TV; Massachusetts Educational Television, Office of Education, WCBV–TV; and independent television producers.

After completing her bachelors degree at Bennington College in Vermont in 1966, she taught elementary school for four years and obtained her masters in education from the University of Vermont (1968). In 1972, she completed her Ph.D. at Fordham University in New York in the field of educational psychology and research with a concentration in social learning and motivation.

Kathleen Devaney is an editor for the Holmes Group, the national consortium of American research universities committed to the reform of their teacher-education programs. She has been a writer, editor, and program manager in education for 25 years. Her experience includes analyzing curriculum programs and directing the Teachers' Centers Exchange at the Far West Laboratory for Educational Development and Research in San Francisco. Devaney's chapter on Family Math is based on her work as a writer and consultant to the EQUALS and Family Math programs at the Lawrence Hall of Science, University of California, Berkeley.

Marvin Druger is professor of biology and science education and chair of the science teaching department at Syracuse University in New York. He has been active in many professional science teaching organizations. He was president of the Association for the Education of Teachers in Science (AETS) and the Society for College Science Teachers (SCST). He was director of the College Division of NSTA and a member of the executive committee. He also served on many other NSTA committees. He served as chair of Section Q (Education) and Council Delegate of the AAAS, and is a member of the AAAS Committee on Council Affairs. He was program director, Science and Mathematics Education Networks Program, for the NSF. He is currently chair of the AETS editorial board of *Science Education.*

Dr. Druger has had extensive experience with radio and television and hosted a public service radio interview program called *Druger's Zoo* for more than 20 years. He also developed a series of television programs concerning careers called *Druger's Working World* for Newchannels Cable Television Network in the Syracuse, New York, area.

Dr. Druger was a Fulbright lecturer at the University of Sydney in Australia. He has received numerous awards for science teaching, including the NSTA 1988 Gustav–Ohaus award for innovations in college science teaching and the first Alumni Award for excellence in teaching from Syracuse University (1987).

Debra Erickson is an education programmer for the San Diego Zoo and has a B.A. in biology from Scripps College and an M.A. in educational technology from San Diego State University. She has been involved in informal science education for the past eight years.

Jon Franklin is a journalist, author, teacher, lecturer, social philosopher, literary theoretician, and the only writer in history to win two inaugural Pulitzer Prizes. Although he has written about a wide range of subjects, he is probably best known for his work on the sciences and their impact on modern life and thinking. Franklin spent 14 years as a science writer for the *Baltimore Sun;* currently he is a professor at the University of Maryland's College of Journalism, where he teaches a variety of courses including reporting, literary journalism, and science writing. He has written five books: *Guinea Pig Doctors* (Morrow), *Not Quite a Miracle* (Doubleday), *Shocktrauma* (St. Martins), *The Molecules of the Mind* (Atheneum), and *Writing for Story* (Atheneum). His latest and most controversial book, *The Molecules of the Mind,* takes a penetrating look at the new science of molecular psychology and the mechanistic view of the human condition being generated by its discovery. A forthcoming book, *America in Amber,* will be published by Simon and Schuster in 1990 and deals with the American culture's fear of the future and the need for a new American dream. He is also working on an introductory psychology textbook, a book on writing, and a book on astronomy.

Andrew Fraknoi is the executive officer of the Astronomical Society of the Pacific and the editor of *Mercury* magazine and *The Universe in the Classroom* newsletter. He teaches astronomy and physics at San Francisco State University and has organized and taught more than a dozen workshops on astronomy teaching in grades 3–12. Books he has written or edited include *Universe in the Classroom, The Universe at Your Fingertips, The Planets,* and *The Universe* (the last two are collections of science articles and science fiction stories published by Bantam Books).

Lynn W. Glass received his Ph.D. in science education from the University of Iowa in 1970. Dr. Glass is a professor of secondary education at Iowa State University. Much of his professional effort in the past decade has focused on developing networks to support science education at the local, state, and national levels. He is director of the Iowa Alliance for Science and the Iowa Junior Academy of Science. Through these organizations he has helped develop and implement several science education projects involving members of both the public and private sectors.

William R. Grant has served as the executive editor of *NOVA* since 1985. From 1983–1985, Grant was the managing editor and series editor for *FRONTLINE,* the public affairs documentary series produced by WGBH–TV. Grant began his career in print journalism, and he served as a reporter and editor for the *Louisville Courier-Journal,* the *Detroit Free Press,* and the *San Francisco Chronicle.* He has published many award-winning articles, and was a Nieman Fellow at Harvard University in 1979.

Pegi S. Harvey is director of education for the San Diego Zoo and Wild Animal Park, where she has supervised educational and interpretive activities for the past eight years. She holds a B.A. with Honors in Education from Arizona State University and a Certificate of Studies from Oxford University in England. She has been involved in zoo education for 15 years.

Don Herbert, of the *Mr. Wizard* television shows, has branched into other areas, including designing science kits on such topics as chemistry, ecology, crystal growing, and electronics. He also has written a series of books featuring experiments that children can do. His latest book, *Mr. Wizard's Supermarket Science,* was published in 1980 by Random House. He has researched, written, and produced 20 classroom films designed to stimulate students' interest in science by having them respond in some way while they watch the films.

Born in Waconia, Minnesota, Herbert spent his childhood in La Crosse, Wisconsin. Trained to be a science teacher at the University of Wisconsin at La Crosse, Herbert instead chose a career in communications that began with six years of acting and writing in Chicago radio. As a captain during World War II, he flew 56 missions as a bomber pilot and was awarded the Air Medal with three Oak Leaf Clusters and the Distinguished Flying Cross.

"Television viewers think of me as a scientist, and scientists consider me a television personality," Herbert explained. "But I'm actually an interpreter who enjoys the challenge of making science and technology enjoyable, appreciated, and understood by the average person of any age. I concentrate on the production, and my wife Norma handles the business side. We enjoy working as a team."

Don Herbert is president of Prism Productions, Inc., and Norma Herbert is executive vice president. They work out of the Mr. Wizard Studio in Bell Canyon, California.

Phyllis Katz left her career as an English teacher in New York City to pursue her interest in language development and the sciences. She created and continues to develop the Hands-On-Science Program with a very dedicated staff in Rockville, Maryland. Her role includes research, writing, and teacher training. She has administered the program's growth from a local PTA experiment to national distribution with the help of National Science Foundation funding. Her husband Victor, a mathematics professor, and their three children sustain the inquiry method during extracurricular hours.

Sol Lander contributed to informal science education by developing "city street" ecology with Biology Regents classes in New York City. After obtaining his bachelor of science degree from Florida–Southern College, Lakeland, Florida, and his masters degree from Brigham Young University in Provo, Utah, Lander embarked on a career of science teaching and supervision in New York City that spanned 33 years. He taught high school biology and general science, then held the positions of assistant principal, biology and general science; director of science, K–12; and assistant examiner, New York City Board of Examiners.

As director of science, Lander promoted training of elementary and junior high school science teachers. He also developed afterschool precollege training for minority students and supervised extracurricular biology clubs. Lander currently serves as a parttime consultant to the New York City Board of Examiners.

Thomas Levenson is an associate producer of *NOVA*. From 1986–1988 he was the science editor for *NOVA*, and before that he was a Macy Fellow in Science Broadcast Journalism at WGBH–TV. Before becoming a television journalist, Levenson worked for *Time* and *Discover* magazines. He published an article on a related subject to the chapter in this book called "Communicating Science, a Media Dilemma" in the January 1988 issue of the journal *Climatic Change*. His first book, *Ice Time: Climate, Science, and Life on Earth*, will be published by Harper & Row in April 1989.

Keith W. Mielke is vice president for research at the Children's Television Workshop (CTW) in New York City. His responsibilities include supervising formative research for all of CTW's educational series, *Sesame Street, 3-2-1 CONTACT*, and *Square One TV*, as well as other projects now under development. Before coming to CTW in 1977, he was a professor at Indiana University's Institute for Communications Research and the Department of Telecommunications. A native of Oklahoma, Dr. Mielke's graduate studies include an M.S. in television and radio at Syracuse University, and a Ph.D. in communication research at Michigan State University.

Lauren B. Resnick is professor of psychology and education and director of the Learning Research and Development Center at the University of Pittsburgh, Pennsylvania. Professor Resnick's primary interest is the emerging field of the cognitive psychology of instruction, and her major current research focuses on mathematics and science learning. She has published widely on these and other topics, including reading, intelligence, and the social functions of testing. She is the founder and editor of *Cognition and Instruction*, a major new journal in the field.

E. Wendy Saul teaches in the Education Department at the University of Maryland, Baltimore County. Her book *Science Fare: An Illustrated Guide & Catalog to Toys, Books and Activities for Kids* (Harper & Row, 1986) was named as one of the outstanding books of the year by the editors of *Booklist*. She is currently completing work on an edited volume entitled *Vital Connections: Children, Science and Books*, which the Library of Congress will publish in 1989.

Edward G. Sherburne, Jr., is director of Science Service, a nonprofit corporation founded in 1921. Its major activities are the publication of *Science News*, a weekly magazine with a circulation of more than 200,000, and a Science Youth Activities program for precollege students, which includes the International Science and Engineering Fair, the Westinghouse Science Talent Search, and the *Directory of Student Science Training Programs for High-Ability Precollege Students.*

Michael Templeton is program director of the National Science Foundation's Informal Science Education program, which supports out-of-school education projects in science and mathematics for both children and adults. The program received more than $13 million in awards in fiscal year 1988. Before joining the NSF, Templeton served successively as director of science at the Pacific Science Center in Seattle, executive director of the Association of Science–Technology Centers in Washington, D.C., and executive director of the Oregon Museum of Science and Industry in Portland, Oregon.

He holds a B.S. in mathematics from Portland State University and an M.S. in physics from the University of Washington. His long-standing interest in science communication in museums and informal settings is reflected in service on board and advisory committees for several national organizations and projects, extensive experience as a museum consultant, and service on the editorial board of the *Museum Studies Journal.*

George Tressel, manager of the National Science Foundation's precollege research and development efforts, is responsible for inschool curriculum and informal learning from such channels as broadcasting, journalism, and personal experience in museums and other activities.

As director of the Public Understanding of Science Program, he helped to develop such programs as *NOVA, 3-2-1 CONTACT, Square One TV, The Brain,* and National Public Radio's science news coverage, as well as *How About,* NSF's successful venture into science news reporting on commercial television.

He played a significant role in establishing the Association of Science–Technology Centers (ASCT), the first organization of science centers, and has instituted innovative programs of cost-sharing, exhibit replication, and traveling exhibits. He is an honorary Fellow of the American Association for the Advancement of Science (AAAS) and ASTC, and chairs the AAAS's Public Understanding of Science and Technology Committee and the Steering Committee of the Fernback Museum of Natural History.

In his many years of film and television production, he has received numerous awards from world film festivals, including Chicago, Atlanta, Edinburgh, and Brussels. In 1964 he directed the U.S. film program for the Atoms-for-Peace Conference and has designed numerous pieces of television and film production equipment.

Strategies

1 A Rationale

George Tressel
National Science Foundation
Washington, D.C.

In a wish-list world, our children would come to school ready and eager to learn, prepared with a background of experience and basic concepts, supported by encouraging parents and friends, and confident that they can learn and succeed if they work hard. Schools would then encourage and reinforce these attitudes and habits, adding skills and knowledge to produce adults who are diligent and prepared for continued learning throughout their lives.

In their earliest years, children should learn to explore the world and how it works, developing self-confidence and zest for thinking and learning. In their middle school years, they should begin to transfer this zest into skills and abilities to articulate questions and to design experiments. Gradually, they should encounter increasing demands in content, skills, problem solving, and integrating and interpreting new knowledge.

However, little of this really takes place. Most

children have almost no formal exposure to science until middle or high school when they have "opportunities" to take subjects like botany, zoology, chemistry, and physics—for which they sometimes have almost no preparation. By age 10 most children decide that science is not for them; it is too "difficult" and "dull." Science has no real meaning for them. It is something that other people do—people who look like Einstein. Science is for men with long hair and white coats who do difficult and dangerous work. Consider these essays by young children (part of the formative research for *3-2-1 CONTACT).*

> I would not like to be a scientist because I would not like to do what they do. They get up early in the morning. I can't get up early in the morning . . . I would not like to be a scientist because I would not like to have to learn all the chemicals. Learn them the way they write them like this: CO_2, H_2O, CO_6. How would you like to write all of them and memorize them too?
>
> Michael, age 11

> I am a scientist. I wake up in the morning very tired. I get dressed. I eat breakfast. I go to the laboratory. I park the car and push a button so my boss can tell I am here. I go to my office and do some paperwork. I have a meeting at 9:30. My friend and I have dug up some dinosaurs' bones. After that I eat lunch and then go to work. After a hard day's work, I go home!
>
> Cynthia, age 8

This discouraging picture need not be. Young children enter school eager and curious about science; if we fail to cultivate these qualities, they change. Children come to dread the demands of taxing formal studies, memorizing facts that seem difficult and irrelevant. They take only science courses required for high school graduation or college entrance.

Obviously, this is not true of all children. Which children go on to succeed? The ones who chose the right parents—parents who bought them books and toys, who took them to museums and encouraged their curiosity, who praised them when they showed zest and talent, who constantly provided stimulus, information, resources, and praise.

Our educational system selects and rewards children who are eager, enthusiastic, and well prepared by their parents *outside of school*. Often, these children are the sons and daughters of the socially and culturally affluent. We must change this. We cannot afford a system that simply selects and cultivates. We must build a well-planned and well-rounded system, both in school and out. Parents are important, but the quality of education and the potential for success should not depend so much on their insight and effort.

We must provide stimulating and rewarding opportunities outside of school for *all* children: television programs that capture the imagination and introduce the exciting puzzles, concepts, and facts of science; museums where children can explore, discover, and experience the joy of learning in a "no fail" environment; books and toys that educate and entertain at the same time; exciting clubs and competitions that prove "I can do it if I work hard." Supported by a consistent and coherent pattern of formal education, a rich environment of informal learning can help make our system truly one of *education* instead of *selection*. If we are successful, all children will share the attitude of this 11-year-old girl:

> I would like to be a scientist who tries to cure diseases. I know it would be very hard and take a lot of work, but I am willing to do it . . . Our science teacher said that all a scientist finds out in his lifetime would take about a billion pages if you put it in a book.

This book is a tribute to the progress we are making in pursuing these goals. The institutions of informal education are some of the brightest assets of American education. Clubs, activities, competitions, and mentors encourage and cultivate the growth of talent and genius. In the past decade, museums, media, and science organizations have made enormous gains. They have become leaders in using recreation for teaching and learning, both by children and adults.

Science on Television

The model established by *Sesame Street* has been developed and replicated in series like *Electric Company, Reading Rainbow, 3-2-1 CONTACT,* and *Square One TV.* Today, substantial numbers of young people watch these series and learn to enjoy reading about science and solving problems. Parents are learning too, as they watch over their children's shoulders. Currently, 50 percent of all 4- to 12-year-olds watch *3-2-1 CONTACT* periodically; 13 percent watch twice a week (Crane, 1987).

Gradually, we are accumulating very encouraging information on the effects of such recreational viewing. Ideally, we would like to see television viewing followed up with direct hands-on experience and activities in places like science museums and clubs. Fortunately, this pattern does happen. More than half of the consistent viewers of *3-2-1 CONTACT* engage in some specific activity afterwards; most often, they go to a science museum with their parents.

Science in Museums and Centers

In the past few years science museums have grown in number, sophistication, and importance. The Museum of Science and Industry in Chicago planted the seeds of this trend at the turn of the century when it emulated the hands-on style and philosophy of the Deutches Museum in Munich. Since then, places like Frank Oppenheimer's San Francisco Exploratorium have raised recreational learning to a new art form. Today, people from around the world look to the United States for examples of the most creative, innovative, and sophisticated examples of informal education.

Almost every science center has a "discovery room" where young children can explore and play scientist. The Boston Children's Museum has carried this concept to a grand scale. The entire museum is planned to deal with children's most important interests and concerns, such as how things work, how things grow, how different people live—even sensitive topics such as dying. Here the serious business of learning is presented as play. Children play together on an "assembly line" to make toys and learn the abstract concepts of cooperation and planning. They learn a tender respect for real American Indian artifacts and the style of life in a Japanese tea room. Or they take home fascinating scrap material donated by local manufacturers, so the experience of experimenting and discovering will continue.

When a museum like the Center of Science and Industry in Columbus, Ohio, invents a new technique, such as a weekend "science camp-in" for Girl Scouts, the idea is instantly copied and embellished. For example, The Lawrence Hall of Science has now developed materials for Boy Scout and Girl Scout leaders to teach outdoor biology. In Philadelphia, the Franklin Institute Museum adapted the concept, and now area teachers bring their sleeping bags for an exciting weekend in the classroom where nobody fails. Perhaps the best complement to this pattern is the flow of discovery-learning techniques into more traditional institutions. Staid and old-fashioned natural history museums—even zoos and aquaria—are beginning to find ways to challenge the visitor to become more involved, and to learn "by accident." Today, a healthy network of talent is working in this area, and new ideas spread overnight.

Science in Clubs and Community

Today, we are extending this pattern through a variety of activity and hobby groups. More than a half-million Girl Scouts have participated in camp-ins and similar programs sponsored by the Columbus Center of Science and Industry and 16 other museums. The Chicago Urban League conducts "Math Counts" competitions to encourage economically disadvantaged students in mathematics. Two years ago, the Girls Clubs of America initiated "Operation S.M.A.R.T." (Science, Math, and Relevant Technology) to encourage girls' interest and ultimately, to integrate math and science into all Girls Clubs programs.

These programs are especially important for women and minorities whose involvement is increasingly critical to our technological future. They are the ones who tend to be ignored, to not have the "right parents," and who are subtly led to expect failure. Too often the expectation is self-fulfilling.

We cannot afford to let this happen. By the year 2000, one-third of the U.S. population will be nonwhite. Even today, the 100 largest school systems enroll almost half of all the minority students in the country, and 23 out of our 25 largest school districts comprise primarily minority students (Association of Science-Technology Centers, 1987).

Lifelong Learning

The changes during the past decade or two are heartening in view of the growing concern about the quality of education. It is encouraging to see growing acceptance of the view that education is a system that includes both inschool and out-of-school experiences. In little more than a decade, we have seen more science on television and radio, in newspapers and magazines, in museums and hobbies. All of these changes mean a richer environment for both adults and children, and a better understanding of the role of science in our lives.

Most importantly, these changes reflect a growing recognition of the importance of informal learning. Words like "hands-on," "experiential," "inquiry," "discovery," and "unintentional learning" have become part of the nation's science education lexicon. The tools and methods of informal education are being recognized as an important component of the educational environment.

Informal learning is important before school, during school, and as a major part of a pattern of lifelong learning. Informal education is a major component of the overall science education pattern—not as a substitute for formal education, but as an important contribution to the preparation for classroom learning and an important influence on enthusiasm, curiosity, and eagerness to study. Equally important, early informal education helps to set a *pattern* of exploring, self-education, and recreational learning.

After all, our ultimate goal is to have a curious, eager, and thoughtful adult. In the end most of us, most of the time, learn most of what we know outside of school. We can all help to build an environment where lifelong learning is commonplace.

References

Association of Science-Technology Centers. (1987). *Involving minorities, women, and disabled persons in science-technology museums:* Washington, DC: Association of Science-Technology Centers.

Crane, Valerie. (1987). *An exploratory study of 3-2-1 CONTACT viewership.* Final report. Chestnut Hill, MA: Research Communications, Ltd.

2 Cognitive Psychology and Science Learning

Lauren B. Resnick and Michelene T. H. Chi
University of Pittsburgh
Pittsburgh, Pennsylvania

For many years now, Piaget's theory of cognitive development has provided the major framework for science educators seeking guidance from psychological research. In recent years, a new wave of research on human cognition has enriched and expanded our understanding of how people of all ages think and learn. Some of this research examines specific forms of science learning and thinking. Together, Piagetian and Post-Piagetian cognitive research provide a framework for designing informal science education activities that will help people learn science.

Piaget's Work

Piaget's work on children's knowledge of scientific phenomena began early in his career. In the 1920s he conducted a series of studies that educators sometimes overlook in focusing on his much better known "stage theory" of cognitive development. In these studies, Piaget (1930) traced the development of children's ideas about natural and mechanical phenomena, such as winds, clouds, floating and sinking boats, bicycles, and steam engines. Piaget concluded that young children do not believe in the necessity of physical causes, but instead believe that things happen because of the intentions and desires of objects. They believe, for example, that clouds move by themselves; the moon follows people as they walk; the pedals of a bicycle do not have to be attached to the wheel to make it turn. Several decades later, Piaget (1974) focused on children's ideas about force, energy, and transmission of movements. He found that young children attributed movements to forces and energies inside objects and believed that people could move objects without touching them directly.

Children also could not coordinate the effects of several variables. For example, when Piaget asked young children to predict the distance that a ball would travel when shot into motion by a spring-powered plunger, their answers were based on only one dimension of the physical situation, such as the size of the ball. Both the early and later studies showed that children gradually outgrow misconceptions and limitations of these kinds as they grow older. They come to assume and search for physical causes; they combine or systematically "factor out" multiple variables; and they no longer attribute intentions and independent actions to inanimate objects. As the children mature from preschoolers to adolescents, their ideas become progressively more like proper scientific ones. Piaget used these and related findings of mathematical and logical performances of children as the empirical foundation of a complex theory of mental development. That theory has two fundamental aspects—constructivism and logical determinism.

Constructivism expresses the idea that people must build their knowledge for themselves. What a person knows is not a simple reflection of what he or she has seen, heard, or been told. Rather, knowledge is a complex result of a long process of personal construction and interpretation. All knowledge, from apparently simple and obvious ideas about number to the most complex theories of motion, energy, and force, must be built through a laborious process in which new and more powerful ideas gradually challenge and replace earlier ones. This is done by *assimilation*, interpreting new information in terms of already established conceptions, and *accommodation*, adjusting and restructuring initial conceptions to take account of challenging data or ideas proposed by others. Assimilation and accommodation coexist in a nearly continuous interplay in all thinking people, resulting in successive stages of *equilibrium*, or states in which people use current conceptions to interpret and explain experience.

Recent research confirms and elaborates on the basic principle of constructivism. Cognitive scientists describe the role of *schemas*—established, organizing mental conceptions—in understanding and learning about new phenomena. Much research shows that what we already know determines what sense we will make of new information—as we read, as we experiment, and as we talk with others. Two major implications of the constructivist principle are that even "low-level" learning involves inferences that go beyond what is actually stated, and subsequent reasoning depends on being able to create "mental models."

Piaget's theory of *logical determinism* claims that as children mature and engage in normal social interchange, they acquire a set of universal logical structures. Although these logical structures are not specific to any domain of knowledge, they determine the kind of reasoning and, therefore, the kind of mental constructions that people will be capable of in any field of science. Piaget claimed that children's prescientific conceptions are primarily a function of the undeveloped logical capacities of children. Their ideas are unscientific because they have not yet developed the general logical structures they need to reason scientifically. According to Piaget, these logical structures develop gradually, in a fixed sequence.

Piaget's three main stages of logical development are well-known: the preoperational, in which children can reason only about physically present relationships; the concrete operational, in which they can imagine certain key actions and effects and thus overcome their dependence on the actual visible environment; and the formal operational, in which they can generate all logically possible combinations and thus engage in fully scientific "logico-deductive" reasoning. Until this logical development is complete, children cannot appreciate or use key scientific processes (such as controlling variables in experiments or combining vectors) and cannot understand complex scientific concepts (such as physical causality). Piaget used this theory of logical determinism to account for the recurrent finding that scientific misconceptions are resistant to training and yet tend to disappear with cognitive maturity. However, five kinds of evidence show that age-related logical development is not an inflexible prerequisite for understanding advanced scientific concepts.

Recent Research

Performance Varies Within a Stage. Many studies show that "stages" of logical development cannot account adequately for the variety of performances seen in children. Children can perform one task (e.g., number conservation) at the level of concrete operations, but still seem preoperational on a number of other conservation tasks (e.g., conservation of weight or mass). Piaget recognized this phenomenon, which he called *decalage*, the French word for time displacement. He explained decalage by proposing that a child can reason operationally when he or she reaches a given level of logical competence, but has to exercise and develop this ability separately for each subject. Piaget thus recognized that specific knowledge plays a role in people's ability to demonstrate logical competence, but he did not admit that knowledge actually helps people to develop their logical competence.

Young Children Reason Logically. Children can also reason operationally well before the Piagetian studies had claimed (Gelman & Baillargeon, 1983). For example, when researchers asked preschool children to decide which of two rows of candies they would like to eat, rather than some abstract (to them) question about whether the number of candies has changed, the children recognized that the number of candies does not increase when the row is spread out. Children recognize that smaller classes of objects (e.g., pine trees) are included in larger classes (e.g., forests) when the wording of the question does not "mislead" them into paying attention to individual objects rather than collections. Children much younger than adolescents can design experiments that systematically vary only one dimension at a time if they know something about the dimensions that are likely to affect an outcome. Furthermore, young children are not necessarily animistic in their judgments of natural and artificial objects, and they reason in terms of sensible physical causality if they have been exposed long enough to a particular physical system to have had time to figure out the causal relations. These findings do not prove that no general cognitive growth occurs, but they do force us to doubt that age-related logical stages strongly limit children's possibilities for learning. They also point to the power of specific knowledge in producing abilities to reason—perhaps even in developing the ability to reason well.

Scientific Misconceptions Persist. Recent research shows that misconceptions about the physical and biological world persist even into adulthood. For example, if college students are asked to draw the trajectory of a ball going off a frictionless cliff at 50 miles per hour, many of them draw trajectories that have either a straight horizontal or a straight vertical component rather than the continuous parabolic curve that would result from the combination of the horizontal and vertical motions. They show the ball moving either horizontally in a straight line off the cliff and only then beginning to fall or curving downward for a while and then dropping straight down. Students explain these drawings by saying that the force causing the horizontal motion eventually dissipates and is "taken over" by gravity, an explanation incompatible with Newtonian inertial theory.

Such misconceptions persist even with excellent formal instruction. For example, many of the students who drew the erroneous trajectories of the ball had taken a formal college course that included Newtonian mechanics. The fact that misconceptions are so persistent suggests strongly that logical development alone cannot ensure good scientific development. On the other hand, the existence of misconceptions shows that people make sense of the world as best they can with the information they have at hand. When their scientific information is incomplete, people develop explanations of their own that may conflict with canonical scientific theories. Once people form such misconceptions, they use them to interpret new information, which makes the misconceptions very resilient. A fascinating example comes from a study by Vosniadou and Brewer (1987), who showed that young children commonly believe that the earth is flat. When they are told that the earth is round, they do not envision it as a round ball, but as a round, flat pancake. The children assimilate the new information that the earth is round into their established idea that it is flat. Studies show that adults also assimilate new information into established conceptions.

Knowledge Changes Scientific Conceptions. Recent research shows that knowledge, rather than logic, changes scientific conceptions. According to this view, children fail to think scientifically not because they cannot reason logically, but because they have not acquired the main organizing principles for some field of knowledge. Carey (1985) showed, for example, that young children initially decide that an object is alive if the object is similar to humans. Researchers asked four-year-olds if an aardvark, bird, worm, cloud, and tools have babies, sleep, breathe, or have bones. Children did not attribute certain properties that all animals share to animals that are very unlike humans, for example, worms. By age 10, however, children knew enough biology to infer more generally that if something is an animal, it must eat, breathe, and reproduce.

Apparently, children learn about biological properties that define life through a mix of formal instruction and informal reasoning. They do not learn other fundamental scientific concepts as easily. To make consistently correct (Newtonian) predictions about projectile motion, for example, children need to undergo a fundamental ontological shift—from believing that motion is a *change* in state and, therefore, must be explained by some cause (such as an external force or an internally stored impetus) to believing that motion is a *state* and, therefore, need not be caused. This is a fundamental principle of inertial physics. Recent research (Ranney, 1987) shows that college students do not induce this principle even when they receive lots of empirical feedback on their incorrect predictions and even generalize to similar situations, but show that they have not induced a general principle by not generalizing to more "distant" problems (e.g., situations where there is no gravity).

Knowledge Produces Reasoning Ability. Knowledge may produce or release reasoning ability rather than the other way around. Children of the same age, but with very different familiarity with a content, are differentially able to reason logically about that content. For example, seven-year-olds who know a lot about dinosaurs infer that a newly presented dinosaur is not a meat eater from the *absence* of certain features. By contrast, less knowledgeable seven-year-olds can only infer that a dinosaur is a meat eater by observing the *presence* of certain features such as sharp teeth (Gobbo & Chi, 1986). Inferring from the absence of a feature is a much more cognitively advanced form of logical reasoning. Children who know more about dinosaurs can reason in that advanced form more than children of the same age who do not have comparable knowledge. Similar studies suggest that how much knowledge a person has strongly affects the quality of reasoning skills.

How to Help People Learn Science

Constructivism tells us that people have to build their own scientific knowledge and understanding and that, at each step in science learning, they have to interpret new knowledge in the context of what they already understand. What people currently understand, however, is often scientifically ill-formed and sometimes even fundamentally at odds with proper scientific conceptions. Informal science education—indeed all science education—must address two paradoxes:

- We cannot teach directly, in the sense of putting fully formed knowledge into people's heads, yet it is our charge to help people construct powerful and scientifically correct interpretations of the world.
- We must take into account learners' existing conceptions, yet at the same time help them to alter fundamentally their scientific misconceptions.

We know that misconceptions are grounded largely in lack of knowledge rather than in failures of logic. We can show how the process of assimilation works. But the question of how to produce *accommodation*—fundamental change in current conceptions—remains a difficult one on which cognitive science has not yet made significant progress. Although we cannot offer clear prescriptions for how to help people construct powerful new scientific conceptions, we can point to a few tested principles that can guide educational efforts. These promising possibilities for new approaches to science learning are particularly suited to informal education settings.

Organize Knowledge. A diet of intriguing facts about this or that scientific phenomenon (as is sometimes offered in films and museum displays) is unlikely to produce the powerful knowledge structures that constitute good science understanding. We know that people simply will not remember isolated facts very easily. To remember, people need to incorporate new information into some organizing structure. In any given field, "experts" (people who already have quite a bit of knowledge about the field) can remember significantly more new information than can "novices" (people with little or no knowledge of the field). An expert chess player can remember a chessboard position much better than a novice player, and children who know a lot about dinosaurs will remember more members of a list of dinosaurs than other children. The reason? Experts can assimilate new information into an existing context. They have a coherent body of knowledge, not merely a collection of interesting separate facts. Lots of facts, even on the same topic, do not automatically organize themselves into productive scientific principles. The knowledgeable person's organized concepts can "explain" and "elaborate" the new information to be learned. To use the dinosaur example again, highly knowledgeable children generate many pieces of meaningful information about a new dinosaur that allows them to classify the new dinosaur. Then they can draw upon what they already know about the class rather than learn a list of separate features.

Those who know more, then, learn more. They learn more easily because they use organizing principles to lighten the new learning load. A challenge for educators is to find ways to help beginners in a field quickly acquire the organizing principles that will "bootstrap" them toward the kind of knowledge-based learning capabilities characteristic of people more expert in the field. Educators must find ways to highlight key organizing principles for beginners and invite beginners to use these principles to interpret particular facts and phenomena. These principles, of course, will be specific to each field and subfield of science.

Making principles explicit for learners probably

helps, but well-organized presentations and explicit statements do not guarantee that every learner will use the ideas exactly as they are given. The educator must think not of providing every detail of a scientific structure for some field of science, but of providing the scaffolding on which learners will be able to elaborate their own versions of a basic structure. This notion of providing the scaffolding is particularly appropriate for informal education settings. Informal educators can take advantage of the special media and display opportunities at their disposal, along with the generally high motivation that people bring to informal environments, to highlight a small number of powerful organizing principles for learners.

Give Time and Repeated Learning. Acquiring an effectively organized body of knowledge takes considerable time. People need to be exposed to an idea more than one time, and each time must elaborate on prior information. People need to have the opportunity and must be encouraged to pry deeply into some topics. A special challenge for informal educators is to provide learners with that opportunity, because informal educators cannot impose a formal curriculum with fixed sequences of activities. Series of films or exhibits on different aspects of the same topic or on closely related topics, with carefully planned adjunct activities, can do much to meet learners' needs. The key is to avoid the temptation to substitute broad "coverage" of topics for depth of treatment.

In this process, providing multiple contexts for any given topic is essential. Exposing learners to a topic more than one time does not guarantee that they will construct organized knowledge. If a topic is repeated but the context is unvaried, a form of *encapsulation* of concepts may occur. Learners will accept new principles or theories but keep them so separate from other conceptions that they have a minor influence on the learners' total mental picture of the world. For example, physics students learn to solve quantitative textbook problems involving Newton's laws correctly, yet behave as if they know nothing about inertial mechanics when asked to think about everyday qualitative physics problems. Educators can counter the effects of encapsulation by showing learners how new principles and concepts relate to many aspects of their existing ideas.

Include Laboratories That Invite and Support Elaboration of Ideas. People learn by personally elaborating on material they are given to study. Students who are passive, whether they are reading a textbook or looking at a lively museum display, do not learn much. Successful learners from physics textbooks, for example, are those who elaborate a lot when they study (Chi et al., in press). In other fields, too, successful students actively relate new, to-be-learned information to their previous knowledge or experiences. Less successful learners, however, are more likely to reread without elaboration when they study (Bransford et al., 1982). In other words, to learn successfully, people must elaborate so that they can incorporate the new materials into their existing knowledge.

Educators can encourage and support the active, elaborative engagement that characterizes good learning. Educators should include questions and problems for learners when they see films, television, or museum exhibits. Museum educators could pose a problem in an introduction to an exhibit, and visitors could then treat the exhibit as a resource for solving the problem. Educators could have their students discuss the exhibit with other visitors and with the museum staff.

Similar activities for informal learning from film and television are essential. Interactive video-based activities might be linked to prepared video materials. Educators can provide laboratories with tools for gathering, reducing, and interpreting data. These ideas are not new to science educators, but cognitive research suggests that they should become increasingly central to informal education efforts.

Promote Discussion. Discussion can help learners to elaborate on their scientific knowledge. People learn better when they discuss and debate ideas with one another. Informal education is ideally suited to capitalize on this natural learning style; people who are captivated by what they see or hear are highly likely to talk among themselves. But just permitting conversation is not enough. Appropriate forms of social interaction need to be "designed into" museum exhibits and into plans for television and other media presentations.

Provide Tools That Help People Construct Mental Models. Much scientific reasoning is more qualitative than quantitative, and scientists and engineers often solve problems as if they were mentally running models of a system in their heads. Many scientific misconceptions arise because of either an absence of a model or a flawed model—one that lacks key constraints and relationships that scientists know about. Simply confronting people with data that contradict their theories is not enough to cause them to reorganize their conceptions. Learners often reject or ignore data that contravene their current views. One reason for this is that, by themselves, learners often cannot construct a mental model of a system that would behave so as to produce the data. Educators can, however, provide learners with the tools for envisioning physical and natural systems and thus developing new and more appropriate mental models.

Film and animation, when presented on computers, can provide models of natural or physical systems that learners can actually manipulate and, therefore, appropriate mentally. Representational tools can help students understand theoretical elements and relationships that they cannot observe directly in nature. Such tools make visible the "hidden structure" of the physical and natural world. For example, White and Horowitz (in press) developed a set of computer-based microworlds for force and motion problems that make vectors and force into physically visible and manipulable objects that behave in accordance with Newtonian theory. The microworlds serve as physical embodiments of the elements in a scientific theory, rather than a display of raw natural objects. These representational tools are powerful because they help people to overcome natural limitations in their ability to visualize and manipulate complex interactions and relations. As a result, they allow people to construct strong mental models of scientific theories.

The Challenge of Informal Education

Within the broad spectrum of science education, informal educators face special challenges but also command special opportunities. In informal education, where a required syllabus and examinations cannot be used to keep people studying, the need for depth is probably the biggest single challenge. Informal educators need to provide depth even at the expense of coverage. Informal science educators should invite people to delve deeply into scientific domains and to build organized bodies of knowledge. Only in this way will people be able to remember what they have encountered in informal science settings and use it to develop further knowledge. Repeated exposure and multiple contexts for exploring any topic are also crucial. Otherwise, the new science learned will not penetrate the main body of learners' ideas.

Informal science education requires far more than attractive exhibits or engaging media presentations. Informal educators must invite learners to solve problems and otherwise go beyond the material presented. Knowledgeable persons must be available to help guide novices as they elaborate. Informal educators need to invite and support social interaction explicitly and not just permit it to happen passively.

The different types of informal education can help learners apply and elaborate their knowledge. Informal settings offer ideal environments for collaborating and debating. Learners can acquire knowledge and ask questions. Media and museum exhibits provide particularly attractive opportunities for using representational and envisioning tools. Educators who supply their students with these kinds of informal learning activities and adapt their teaching methods accordingly will greatly enhance the effectiveness of their work.

References

Bransford, J.D., Stein, B.S., Vye, N.J., Franks, J.J., Auble, P.M., Mezynski, K.J., & Perfetto, G.A. (1982). Differences in approaches to learning: An overview. *Journal of Experimental Psychology: General, 111*, 390–398.

Carey, S. (1985). Are children fundamentally different kinds of thinkers and learners than adults? In S.F. Chipman, J. Siegel, & R. Glaser (Eds.), *Thinking and learning skills: Vol. 2. Research and open questions* (pp. 485–517). Hillsdale, NJ: Erlbaum.

Chi, M.T.H., Bassok, M., Lewis, M., Reimann, P., & Glaser, R. (in press). *Self explanations: How students study and use examples in learning to solve problems.*

Gelman, R., & Baillargeon, R. (1983). A review of some Piagetian concepts. In J.H. Flavell & E.M. Markman (Eds.), *Handbook of child psychology: Vol III. Cognitive development* (pp. 167–230). New York: Wiley.

Gobbo, C., & Chi, M.T.H. (1986). How knowledge is structured and used by expert and novice children. *Cognitive Development, 1*, 221–237.

Piaget, J. (1930). *The child's conception of physical causality.* London: Routledge & Kegan Paul.

Piaget, J. (1974). *Understanding causality.* New York: Norton.

Ranney, M. (1987). *Restructuring naive conceptions of motion.* Unpublished doctoral dissertation, University of Pittsburgh, Pittsburgh.

Vosniadou, S., & Brewer, W.F. (1987). Theories of knowledge restructuring in development. *Review of Educational Research, 57*, 51–68.

White, B., & Horowitz, P. (in press). Thinker tools: Causal models, conceptual change, and science education. *Cognition and Instruction.*

3 Challenges in Audience Research for Informal Learning

Valerie Crane
Research Communications, Ltd.
Chestnut Hill, Massachusetts

Visiting a museum exhibit, watching a local newscast, viewing a talk show or public television program, and listening to a radio program are all informal learning activities. Although varied, informal learning activities have a common goal: to communicate specific information to targeted audiences in an entertaining or interesting way. Many alternatives compete for the public's leisure time, making the job of producing entertaining educational programs or activities difficult and often expensive. Research Communications Ltd. (RCL) was established in 1980 to help producers of informal learning activities attract target audiences and increase the audience's appreciation and understanding of the messages conveyed. RCL uses a range of audiences, including preschoolers, elementary age children, teenagers, and adults, to evaluate the entertainment and interest value of informal learning activities. Some projects require us to analyze audiences in a single local market; other projects have nationwide or worldwide audiences.

The method of audience analysis combines quantitative and qualitative research methods to provide project developers with diagnostic information about the program or exhibit being tested; data that indicate performance potential after the program or exhibit is broadcast or shown; and comparisons of reactions among different audience segments.

Because producing a television program or developing a museum exhibit is relatively expensive, we often conduct formative research in the development process to assist the design team. Formative research involves ongoing testing of content and prototype materials as they are created. If a wrong step is detected midway through the project, it is much easier to correct than after the project is complete.

Formative research in television was first conducted at the Children's Television Workshop in the late 1960s. Their ongoing testing of segments shaped the development of *Sesame Street*, a legendary success in informal learning. Children's Television Workshop established an inhouse team of researchers who performed the audience research each project required for development. However, most organizations cannot support an inhouse team of researchers in the way Children's Television Workshop did, because many production efforts are not ongoing and therefore cannot support a fulltime, inhouse research staff. This is where independent companies that provide audience research play a role in informal education.

The target audience and the learning activity are not the only determinants of what gets produced in informal education. The source of financial support also plays a role. In the past decade, sources of public money have shifted their priorities from content to audience. This shift means that financial supporters are not concerned with merely "correct" content. They want to increase the public's interest, appreciation, and understanding. While this shift might seem subtle, it is dramatic because it affects how informal learning activities are designed and produced. As the public becomes the focus of informal learning, how to communicate content effectively becomes increasingly important, particularly during the early stages of project development. Designers of informal learning activities not only need to be concerned that they send accurate messages to the audience, but also that the messages are received, understood, and interesting to the audience.

Involving the audience in the development of informal learning experiences may sound like a laudable goal, but an obvious question is "How are the needs of the project and audience met?"

Challenges in Evaluating Informal Learning

Audience research for television and museum projects presents unique challenges. The first task in research is to define the target audience for the informal learning activity. This includes targeting demographics such as age, gender, education, and income, as well as relevant background information such as level of news viewing or frequency of museum visits. The ability to generalize the data will be based, in part, on how representative the audience is of the actual targeted audience for the material. Testing prototype science exhibit material with individuals who report they would never visit a science museum will render useless information. Testing a morning talk show with viewers who are only available to view television in the evening also has obvious problems. Evaluating a science series with students enrolled in science courses will provide good information about the potential use of the series for formal learning, but unreliable information about potential for home viewers who are not enrolled in college-level science courses.

The second step is to convert specific objectives into a set of questions that reveal the interest and informative value of the materials being tested. Questions can be divided into two general areas.

The first is the affective area—determining interest and attitudes toward the subject as well as satisfaction with the materials. The second important area relates to cognitive variables that include prior knowledge of the subject and learning outcomes. Both affective and cognitive variables include perceptions of informative value and expectations for and satisfaction with learning.

Our research shows that prior interest in the subject is one of the most powerful predictors of participating in informal learning activities. While television producers, news directors, and exhibit designers must not limit themselves to those areas in which the public is highly interested, they should know levels of interest for two reasons. First, the public wants information that interests them and that will help them understand their world. It behooves informal educators to serve those needs. On the other hand, sometimes news stories must be told, and unfamiliar and difficult science content must be presented as part of educating the public. It is even more important to understand *why* the public lacks interest in less compelling subjects, and to find ways to help the public understand and become interested in the subjects.

Attitudes about subject matter also are important. On one project, we measured how the audience felt about aging. Using a series of agree disagree statements, we found that the audience held certain misconceptions about aging. Since the misconceptions seemed fairly straightforward, the design team presented personal stories that countered the misconceptions. However, the misconceptions persisted even after audience members viewed the program. The design team then added more explicit statements to counter the misconceptions and found that the stereotypes were effectively overcome.

Another area related to the affective domain of informal learning is satisfaction with material. For example, by using slides that depict different elements in an exhibit, we can effectively measure visitors' satisfaction by tracking their responses. We can measure specific elements within the learning activities, such as types of exhibit displays or a host of a television series. These factors relate significantly to overall satisfaction with informal learning activities and, hence, the likelihood that audiences will participate in them.

In the cognitive area, we look at prior knowledge of the subject matter, comprehensibility of the materials, and perceptions of the informative value. Looking at knowledge of the subject is important in designing materials at the appropriate level. If the learner's knowledge does not match the targeted level of the materials, the informal learning activity risks failure.

Comprehensibility may or may not be strongly related to subsequent participation in informal learning. If informal learners report that the materials are uninformative or incomprehensible, the likelihood of sampling a program or exhibit is greatly diminished. However, if a comprehension test item reveals that the exhibit or program actually misinforms the audience, but the audience does not realize they're being misinformed, the test item does not effectively predict overall satisfaction.

However, learning perceptions and informative value correlate strongly with a learner's continuing participation. If viewers rate a newscast or museum exhibit highly for its informative value, they are more likely to view that newscast again or return to the museum.

The greatest limitation in conducting formative research on programming and exhibits is in discovering how much long-term learning has been acquired. We cannot predict what audience members have learned from the entire series or finished exhibits that have limited exposure. However, we can demonstrate that cognitive learning occurs from a single exposure, and designers should not be reluctant to demonstrate this learning, however limited. Equally important, we must keep in mind that the goals for the informal learning activity may differ markedly from formal learning in this regard. The activity may not result in higher standardized test scores, but it will spark learners' interest in a subject and increase their awareness and appreciation.

Several methods are used to measure audience response to informal learning. Telephone surveys are probably the most common approach to gathering quantitative data on large-scale samples of the public. They can provide tracking information on attitudes, preferences, and actual habits and behavior. However, when testing materials in the developmental stages, it is necessary to bring groups of the target audience together to test the materials. News viewers can recall very little specific information over the telephone, even after viewing a particular newscast. A sensitive and commonly used research method is to have a group watch a newscast, answer questionnaires, and participate in indepth interviews. This method shows the diversity of an audience and, at the same time, shows trends that are useful for decision-making as materials are developed. Critics of the method claim that the group environment differs too greatly from the natural informal learning environment, that group discussions contaminate individual responses, and that they cannot predict how well a program or exhibit will perform once completed.

A recent study of a museum exhibit illustrates how predictive group testing can be used. Our researchers used a series of slides of the actual exhibit to test visitors' satisfaction. The researchers sampled visitors as they left the exhibit, which was considered successful because of high attendance. To determine whether or not potential visitors would respond in a similar manner, researchers used this set of slides at the next location of the exhibit. On all but 2 out of 25 slides, the potential visitors' interest levels were comparable to those actually visiting the exhibit. In the group discussion, the potential visitors cited many of the same criticisms, both positive and negative, about the exhibit as had the actual visitors. These findings suggest that the "formative" methodology of showing only slides before visitors see an exhibit can be a very effective tool for determining how to refine traveling exhibits.

Does Audience Research Make a Difference?

In the past, many educators considered informal education to be a trivial undertaking compared with the challenge of providing high-quality formal education. But in today's changing environment, particularly in the field of science, informal learning activities actually occupy a greater place in our lives because we continue to learn informally long after our formal education has ended.

Another destructive attitude that emerges in discussions with both formal and informal educators is a disdainful attitude about the public. Project developers comment often on a poorly educated public, who, they say, can't decide for themselves what is best for them. This elitist view has prevented educators from using audience research for informal (or formal) learning. Perhaps one of the most positive, unexpected outcomes of our research efforts occurs when a project team observes a group test session and realizes that the public can talk intelligently about how well informal learning activities meet their interests and needs.

But the challenges for providing systematic informal learning are great. When we talk about formal learning we are talking about relatively targeted populations. When we talk about informal learning we are talking about unlimited target audiences. The challenge is to serve a range of audiences effectively.

Formative research brings the public into the process, and serving the public is what informal education is all about. Increasing the public's appreciation and understanding of science is possible only if the intended message reaches the public. And the intended message will only reach that audience if it captures their attention, maintains their interest, provides a compelling experience, and fosters satisfaction. Audience research is merely a roadmap toward these goals.

Helpful Hints

- Begin by setting goals for your informal science education activity. That activity may be a club, workshop, television program assigned as homework, or a field trip to a museum or zoo.

- Develop a simple evaluation tool to provide feedback that will determine whether or not you have met your goals. The following quantitative and qualitative questions may help you evaluate the activity and teach your students to evaluate their own experiences.

  How would they rate the experience on a scale of 0 to 10, where 0 is the lowest and 10 the highest?
  Why did they give it that rating?
  What did they like best about the experience?
  What did they like least about the experience?
  Did they feel that they learned something?
  What did they learn from the experience?
  How clearly was the information presented?
  How might the information have been presented more effectively for the target audience?
  Would they repeat the experience (watch another program in the series, visit the museum again on their own)?
  Why or why not?

- Process this information by computing simple frequencies and percentages for the quantitative question and compiling the open-ended responses.

- Share and discuss the findings with your class. Consider having the students prepare a simple report on their findings, and either send the reports to the producers of the television program or to the museum.

- Keep a record of the different activities that you schedule from semester to semester and year to year to see how effectively you provide informal science learning as part of your overall science education program.

The Media

4 NOVA, Television, and Informal Science Education

Thomas Levenson, William R. Grant, and Paula Apsell
WGBH–TV
Boston, Massachusetts

The first task of any prime-time television show is to entertain; if it founders there, it will fail any other mission. *NOVA* is fully aware of that fact, and has chosen to complicate its mission by opting not only to entertain but to educate.

In fact, that double imperative defines *NOVA*'s mission and marks its usefulness as a vehicle for educating the American public about science. We reach an average of 11–12 million people each week, which pales by comparison with the audiences of top-rated commercial television series like *Dallas* and *The Cosby Show*. But our audience is large in public television terms, rivaling that of dramatic shows like *Masterpiece Theater*. Demographic studies of *NOVA*'s audience reveal they are somewhat older, better educated, and better off economically than the nation as a whole. *NOVA* has an audience of curious grown-ups rather than the stereotypical, average television viewers.

Telling a Story

Such an audience confronts *NOVA* with several programming problems. We must ensure that each program and the mix of programs in a season contain enough startling information and visual images to capture the attention of 12 million viewers. We also must ensure that the scientists concerned with each program have no chance or cause to complain about our treatment of their ideas. We don't always succeed. In a 1985 show that covered the especially troublesome subject of genetic manipulation's potential to prevent or cure inherited diseases, the technical accuracy of the facts was much more apparent than the entertainment value of the program. By contrast, from the same season, the program "The Case of the Frozen Addict" dealt with an equally difficult subject—it described the progress in understanding Parkinson's disease. We feel that it was a far more successful film. The reason? It was a good story, interestingly shot and well told.

We learn and relearn with each project that the *subject* does not make *NOVA* work, the *story* does—a discovery that provides endless frustration for our scientist/advisors. Recently, a major chemical journal published a letter complaining that *NOVA* produced shows on chemistry far more rarely than it did on physics. That's true, but it misses the point. We don't see ourselves doing "physics" programs or "chemistry" lessons. We see ourselves doing stories that demand an explanation of the science involved in order to make sense.

For example, *NOVA*'s film "Sail Wars" was not specifically about physics. But it did succeed in demonstrating some basic but difficult physics principles while describing how researchers design world-class racing yachts. This film was among the best *NOVA* has produced over recent years. Our confidence in making that judgment illustrates another fact of broadcasting informative shows to the public: Because we are a public broadcasting endeavor, we have a certain cushion or leeway that enables us to use intelligence in a way that isn't possible on the commercial networks. Public broadcasting lives by the ratings, but in a far more indirect way than any commercial effort. A program on PBS that attracts no audience will eventually die, but it will usually have time to prove itself. This means that ultimately we make decisions based on each program's merits—the years have shown we can rely on our audience to give us the benefit of the doubt.

That way lies hubris. However, it is startling that our judgments about what is good and bad are almost entirely unrelated to the ratings we gain on each program. It is not that we think unpopular shows are good and popular ones are bad; we just can't find any consistent pattern when we eliminate the most popular programs from the discussion. For example, our complicated genetics program, despite all our efforts, was seen by four percent of the people watching television at that time—slightly worse than the average *NOVA* share of five percent of the national audience. A 1985 robotics program drew a slightly above average six percent of the viewing audience, and yet ranks in our minds as one of our less successful efforts. With a slightly more positive outlook, we cite the first film of the 1986 season, "The Search for the Disappeared," a program about forensic medicine and genetic screening at work in postdictatorship Argentina, which drew a four percent share of the audience. "High Tech Babies," another excellent film from the 1986 season, did the same.

Playing the numbers game, of course, is a risky business. Interpretation is always a problem. "The Search for the Disappeared" aired opposite the baseball pennant playoffs. Does a four percent share represent a falling off or a triumph in the face of overwhelming competition? We don't know. The second highest *NOVA* of all times, "Nomads of the Rainforest," drew 10 percent of the audience when it aired against the three commercial networks' report of Ronald Reagan's landslide victory over Walter Mondale, a story that had lost its cliffhanger element by our 8 p.m. airtime.

But what about that handful of *NOVA* programs that attracts an unusually large audience? These programs, like "Nomads," tend to be aired at an especially fortunate moment, or to be films whose subject matter alone attracted an audience much larger than *NOVA*'s regular set of viewers. The highest rated *NOVA*, the 1985 "Case of the Bermuda Triangle," attracted slightly more than 11 percent of the national audience. Generally, our natural history and wildlife films also perform above the *NOVA* average; "The Crucible of Life," a film about the natural world within a Hawaiian volcanic region, gained 10.7 percent of the national audience on its broadcast date.

The imperative would be to craft a *NOVA* season of programs about sharks, whales, volcanoes, and strange stars and planets. Clearly we have resisted this temptation, which leaves us with the uncomfortable sense of being unable to use the ordinary tools of television programmers as our guide. So we have tried to come up with a set of criteria of our own. These criteria generate a series that performs something of an educational role; the decisions we make on each film, however, do not begin with any sense of the film's value as pedagogy. First, we look for something that will be a good movie.

To a certain extent, that judgment is dictated by the demands of our format. The definition of what our format requires is also a definition of what our format prohibits. A *NOVA* story has to hinge on science, somehow or other—although we define science as broadly as possible. The story must fill an hour. It must have a strong visual component, or failing that, a producer who can transform an apparently unvisual story into a tale in pictures. (The best example of this technique is a wonderful film we coproduced with the BBC, "It's About Time." In this film, Dudley Moore encounters St. Augustine and the mysteries of jazz piano while trying to explain the extraordinarily slippery concept of time. You can't see time, but the program worked.)

Choosing a Topic

Happily, some stories that fall out of *NOVA*'s format can be done in other venues—especially such programs as *Newton's Apple* or *3-2-1 CONTACT*. But U.S. television does not have a science magazine program for curious grown-ups that breaks news in the sciences as effectively as *Nightline* deals with public policy issues or *Entertainment Tonight* deals with show business. This lack leads to the next imperative for *NOVA*: to choose each season's mix of shows with some sense of the journalistic demands of covering contemporary advances of science.

For some time, we have discussed the need to do a film about quantum mechanics—the field of physics that introduced the notion of probability to understanding the material universe. The discussion has always chilled our collective bones. We have come up with sketches for films on quantum mechanics and changing notions of reality; quantum mechanics and applications in other sciences, especially chemistry, materials physics, and some arcane biochemistry; quantum mechanics and the structure of matter, and so on. All are subjects, not stories. Not one has an obvious visual hook. All hinge on science, but that alone can't make a *NOVA*. As for length, any of these could expand to an hour or longer; our hapless producers could try to explain arcane concept after concept. And yet, America's only science documentary series cannot forever ignore the most fertile theoretical tradition in what is considered the queen of the sciences. What to do?

In this case, turn to the distinguished Scandinavian film industry. By the time this book appears, we will have either completed or nixed a *NOVA* that originated as a Danish film about the Einstein-Bohr debate over one of the core predictions of Bohr's quantum theory. The Danish producers used some graceful techniques, such as still photographs, to tell the story of Einstein's distress at the idea of a universe governed by probabilities. The film counters with Bohr's belief that reality is a relationship between the observer and the world at large, and shows Einstein's efforts to come up with a paradox that would doom Bohr's construction. It concludes with a description of the recently completed experiment that demolished Einstein's paradox and confirmed Bohr's theory, and hence, the validity of quantum mechanics.

A number of cinematic features attracted us to this film. The Danish producers mixed photographs of Einstein, Bohr, and their circle of colleagues with elegantly shot and conceived interviews of men involved in the debate. The result is a film that gives a sense of Einstein and Bohr moving through the years to their ultimate argument, without any actual moving footage of the two scientists. But more important, this film meets the main criterion for success as a *NOVA*. It is a clearly defined story that examines the largest intellectual concerns in the history of twentieth-century physics.

Finding Perspective; Giving Perspective

Assuming we are able to adapt the Danish work to our needs, the film will, for a time, ease the pressure to produce a theoretical physics film. It isn't just that certain fields in science keep evolving and need periodic re-examination. Science is embedded in a host of public issues, from environmental ones like toxic waste disposal to problems of national defense. Perhaps the most compelling example is Ronald Reagan's proposed Strategic Defense Initiative (SDI), commonly known as Star Wars. In the popular media, the debate seems to be about whether or not to build the system. Impressive drawings of space-based weapons show what the system could look like, and elaborate schema have been proposed to show how such a defense could be organized; in effect, the argument seems to be a reprise of a hundred other large procurement debates.

In fact, the argument is a scientific and a technological one that is confused but not resolved by the politics of appropriations on Capitol Hill. On one hand, a large body of American scientists argues that Reagan's vision of strategic defense is impossible or unrealizable for reasons fundamental to the physical principals involved and because of facts of technological life. On the other hand, a smaller but quite expert group of scientists, mostly in national labs, concedes that SDI is impossible now but that given enough money, the group can figure out the whole problem.

NOVA joined the *Frontline* series to produce a two-hour film on SDI to address this debate. SDI raised a dilemma for us. Although SDI is a technical project and should be analyzed on its merits, it is perhaps the most significant political debate of the last few years. SDI, after all, is Reagan's answer to critics who claim his stand on arms control has made the United States less safe than it was before he came on watch. Even a perception of criticism may appear as a direct attack on national policies, and *NOVA* has no particular desire or mandate to produce that kind of editorial.

Yet SDI is the largest scientific and engineering initiative mounted by the U.S. government, and it raises a number of questions that are clearly within our ordinary set of interests. Ultimately, we conceived the mission of our film as one that raises the level of discourse on the subject—that enables people to evaluate specific claims for SDI but that won't be seen as an attack on American security. Explicitly, we hoped to raise the level of competence in the political press about SDI, as well as in the larger audience that watches us and reads the newspapers. The film described how proposed weapons function; why they might or might not be able to fulfill proposed missions; how systems function; what particular problems result as the scale of a complex system grows; and finally, whether or not countermeasures to proposed devices and systems pose a major threat to Reagan's proposal. We hoped that both the public and the press would use this beginner's guide to strategic defense to evaluate claims made by either side.

Call this a straightforward effort at public education if you will. As such, one has to acknowledge that it was pretty much a failure. Even a cursory read of press reports of arms control negotiations indicates that diplomatic and White House reporters continue to portray SDI as a sort of scientific magic shield. They only ask if it is diplomatically wise, not whether it is even scientifically possible.

Given that we occasionally produce an almost purely educational film, and that we don't feel these films have a great impact, how do we see our role in bringing science to the American public? We like to think that an entire season or more of *NOVA* represents a fairly high level of science education and that as a result, the scientific literacy of our audience steadily increases. But even if we could test this thesis, it's clear that our educational impact is incremental at best. The important issue is not that our audience retains specific facts from any given program. Instead, they should understand that science is a human endeavor that we use to order our material world. If our audience gets a sense of the method and power of science and the limits to that power, then we have done a good job.

We find ourselves in the same predicament as the nightly news. We write "headlines" about what is important in science in a given year; to really get a handle on the whole story a viewer must go well beyond what we can give them. And so, we return to where we began: Successful television has to recognize that the medium is better suited to entertainment than to pure pedagogy. We can arouse someone's curiosity, but the rest is up to the viewer, and perhaps to you.

5 Making Informal Contact

Keith W. Mielke
Children's Television Workshop
New York, N.Y.

Picture yourself walking into an elevator to find, to your surprise, a young man standing on a scale, weighing himself as the elevator moves up and down. You are intrigued. You wonder where he got the idea to try this—and why? A Field Producer for *3-2-1 CONTACT* actually found herself in this situation. Although suspecting the impetus behind the youngster's activity, she asked him anyway. He said that he had just watched an episode of *3-2-1 CONTACT* in which Paco, one of our cohosts, modeled this activity and pointed out scientific principles pertaining to gravitational forces and weight.

3-2-1 CONTACT, a science series produced by Children's Television Workshop for 8- to-12-year-olds, airs daily on most PBS stations. We continually encourage our viewers to try out experiments on their own, and whenever we hear anecdotes like "the elevator story," we rejoice, for this anecdote is yet another testimonial to the power of television used in the service of informal science education.

Television's Potential for Informal Science Education

What can we offer in the way of science education for young people outside the classroom?

"Outside" the classroom does not mean "in competition with" the classroom, or a "substitute" for what teachers can do. The goal is to complement the classroom experience. The face-to-face environment of the classroom takes maximum advantage of what the teacher can do within the immediate range of the senses. A complementary form of education with television offers something the teacher cannot, such as a "field trip" to a rubber factory in Malaysia or a "chat" with a famous scientist. It is not necessary to build the case for one medium such as television at the expense of some other medium. *All* media can do some things better than others, and so it makes sense to use various media in concert, exploiting the strong features of each.

For *3-2-1 CONTACT*, informal education means that success is not defined as some single, tightly defined outcome; instead, successful outcomes can take many forms, including, as noted previously, taking the bathroom scales for an elevator ride. Not everything has to be explained until comprehension has reached some criterion level. People bring enormous differences in background and education to the screen. We do not attempt to isolate subcategories of content in a structured and sequential way. On the contrary, we attempt to portray science in a broad context—as part of life.

Love of science starts with activities that are fun and phenomena that are interesting. Novelty, wonder, and imagining "what if" can stimulate young curiosities. After this "hook," the intrinsic reward and motivation of formal study can take over. From interviews with scientists, we know the effects of informal science education are not restricted to semester-length periods, but instead develop over many years.

For example, we interviewed a scientist born just after the turn of the century. He recalls vividly his first informal lesson in applied mathematics as a young boy in a small Oregon town. He had just received a jackknife as a present, and was trying unsuccessfully to sharpen a pencil with it. A cowboy sauntered by and stopped to explain a principle: If you first establish an angle of cut, and then follow that angle as you rotate the pencil, you can neatly sharpen the pencil. The name of the cowboy is lost to history, but the name of the young boy is Linus Pauling, and the effect of the informal education incident was long-lived and fruitful.

Another scientist *3-2-1 CONTACT* interviewed recalls that as a young boy he noticed a very large book in the library with the simple title *Heat*. How, the boy wondered, could anyone possibly write such a large book about such a simple subject? Things are either hot or cold, and that's about it. The boy opened the book, wondering what it could possibly contain. He saw a symbol he didn't understand, but recognized it as some form of high-falutin' mathematics. A prescient insight occurred: If there was this much to know about heat, and if higher mathematics was needed to describe it, then an entire set of relationships existed that he had never suspected, and higher mathematics would be useful for understanding lots of things. You may protest that this is not your typical boy in your typical library, and you may be right—his name is Steven Weinberg, who now is a Nobel Laureate in physics.

But what about the many, instead of the few, and what does a mass medium such as television have to offer? Some people have trouble thinking of television in the same category as serious education. Those who concentrate only on the "dark side" of television—violence, passivity, escapism, superficiality, and the like—can easily find plenty of examples to illustrate their case, of course. The other side, however, is the positive and constructive use of the same powerful medium.

What are some of the positive features of television? First and foremost, kids like television and watch it a lot. When they aren't in school, where formal science education takes place, television excels in reaching kids at home, where there's enormous opportunity for informal education. Children watch about 25 hours a week, and by high school graduation, they will have spent more time watching television than attending classes. Regardless of one's personal opinions about television, these powerful statistics are a fact of life and an educational opportunity to be reckoned with.

Television is well suited in other ways to serve many needs of informal science education. Television is a democratic technology, offering identical programming to all children, regardless of where they live or how much money the local school can afford for science instruction. We know classroom opportunities are not equal across the country, but essentially, access to television is. Television is superb for telling stories and communicating the human dimension of feelings. Television can induce great mental involvement. In the hands of a skilled production team, television can stimulate and enhance thought processes with visuals and graphics. Through its ability to provide "eyes-on" experience, television expands a child's world beyond the hands-on, immediate level.

Television is an extremely information-dense medium—more content is presented than received and interpreted. Different people extract different meanings and information from it, and more information always remains. The content of a book about volcanoes, for example, can presumably be defined as its verbal content. However, a television segment on volcanoes, such as *3-2-1 CONTACT* featured, is endlessly rich in what can be perceived, because of its verbal *and* nonverbal content. Do the shades of color in the lava flow correlate with different temperatures? How hot was the lava when a tool, accidentally dropped, burst into flame? How do you feel as the helicopter pulls away, leaving two people close to an active eruption? A serious attempt to describe the content of the segment would fill many pages, take a very long time, and ultimately fall short of being complete. This richness of content in television programming serves diverse audiences particularly well when it is applied to informal science education. Very young children extract one level of meaning, older children another, adults another, and professional scientists yet another.

Television's home environment presents unique demands along with its opportunities for teaching and learning. The audience must be attracted and entertained as well as taught, for turning off the set is just as easy as turning it on. No rule states that one must watch television; other activities vie for children's time and attention as well. Viewing is voluntary. Audience members must choose to turn the dial to *3-2-1 CONTACT.* The series' creative team of producers, content experts, and child researchers combine talents to produce a program that attracts attention, educates, and entertains.

What we deal with ultimately, however, is not an abstract theory of what television can do, but a real-world application of what a team of producers, researchers, and content specialists can do, and have done, in a specific case, such as *3-2-1 CONTACT.*

3-2-1 CONTACT

3-2-1 CONTACT uses the power of television. Designed primarily for home viewing and secondarily for school and community use, the series has three major purposes:

- to help children experience the joy of scientific exploration and creativity and the satisfaction of accomplishment; to motivate them to pursue further scientific activities;
- to help children become familiar with scientific thinking and to stimulate their own thinking skills so that as they grow, they can analyze important social issues related to science and technology; and
- to help children, especially girls and minorities, recognize science and technology as a cooperative endeavor, relevant and meaningful to their lives and open to their participation.

The series is designed to function as informal education, but that does not imply the content is assembled casually. A careful planning process is conducted for all shows, which are organized into weekly themes, to ensure the result is good television, good science, and good programming in terms of appeal and comprehensibility for our target audience, 8- to 12-year-old kids. The day-to-day management of the project reflects these priorities with coordination among the departments responsible: Production, Content, and Research.

Some theme weeks, such as "Stuff" (materials and their properties), "Flight," "Signals," and "Water," reflect broad categories of phenomena. Others, such as "The Tropics" or "Japan," are organized around geographic locations. Each theme is featured in five related shows broadcast daily after school. (Many stations also make the series available for inschool use by airing the series during school hours.) Each program is designed to stand alone, but children who watch several programs within a theme week should gain a more detailed perspective on the programs' unifying scientific principles.

3-2-1 CONTACT's magazine format combines live-action sequences, on-location documentaries, animation, and stock footage assemblages. Many programs contain a mini-mystery serial (an audience favorite) that features the adventures of The Bloodhound Gang, whose protagonists solve mysteries through scientific reasoning and knowledge.

From inhouse interviews and studies we have conducted with target-audience children, we know that viewers retain a large amount of information presented on the show. Furthermore, an NSF-sponsored study of how *3-2-1 CONTACT* functions as informal science education found that more than half of the viewers went on to *do* something as a result of watching the show.

The longer term goals and the idiosyncratic effects are more difficult to assess, understandably, particularly with quantitative measurements. We therefore rely a great deal on guidance and feedback from our National Advisory Committee, composed of distinguished scientists, science educators, and public broadcasters. We also find good anecdotes to be rewarding and informative, regardless of their scientific status. For example, one of our young viewers was depressed by the news that he needed heart surgery, specifically the implantation of a pacemaker. However, after coincidentally viewing an episode of *3-2-1 CONTACT* that presented and discussed pacemakers, both the child and his parents were relieved and uplifted. Sometimes children write to us, suggesting their own story ideas. One child asked for a show on interspecies communication, and an entire fifth-grade class requested a show about silkworms and the production of silk thread.

Nor are these anecdotes restricted to the United States. Over the years, *3-2-1 CONTACT* has been broadcast in 33 foreign countries, including technologically advanced and underdeveloped areas. Professor Mary Budd Rowe, *3-2-1 CONTACT* Advisory Committee Member, once spotted a group of children running excitedly down the street in Katmandu, Nepal. Curious about their destination and purpose, she followed them to a storefront, where the youngsters went to view *3-2-1 CONTACT*. Similarly, two of our *3-2-1 CONTACT* inhouse staff members were in Malaysia to scout story ideas. They were in a rather remote area, and stopped to ask directions. There, a mother and nine children, sitting on the dirt floor of their modest hut, were viewing *3-2-1 CONTACT* on their small black-and-white television.

One young viewer paid us what might have been his ultimate compliment: He said he likes *3-2-1 CONTACT* "even better than cartoons."

Other Uses of *3-2-1 CONTACT*

So far, we have been discussing the informal use of *3-2-1 CONTACT* only for voluntary home viewing. However, other opportunities exist outside the school setting. Community Education Services (CES) is the Children's Television Workshop department that specializes in this category.

In cooperation with concerned adults and financial sponsors, CES has organized hundreds of *3-2-1 CONTACT* Science Clubs around the country. These clubs usually meet at schools, but they are not part of the formal science curriculum. A typical club consists of approximately 10 youngsters and an adult leader who use *3-2-1 CONTACT*'s weekly themes to generate ideas and to structure activities. For example, during "Water" week, club members might perform the "Three Faces of Water" experiment that shows how water looks as a solid, liquid, and gas. They might also set up an aquarium, make a water wheel, or visit a local weather bureau office. These activities motivate children and help them find, understand, and appreciate science and technology in their daily lives. Science museums, the Girl Scouts of America (which uses *3-2-1 CONTACT*-based activities as part of its merit badge program), Girl's Clubs, Boy's Clubs, 4-H Clubs, and YMCAs have all sponsored *3-2-1 CONTACT* clubs.

3-2-1 CONTACT: The Magazine

3-2-1 CONTACT is also in print: 375,000 households subscribe to the award-winning *3-2-1 CONTACT* magazine. The magazine contains articles on current science-related topics, as well as games, puzzles, and activities to motivate and interest 8- to 14-year-olds. In addition, it has a special section that concentrates on computer-related subjects.

Although issues of *3-2-1 CONTACT* magazine are not tied to specific *3-2-1 CONTACT* shows, the magazine's philosophy parallels that of the series and, therefore, reading the magazine complements viewing the show.

All Work Together

Television can be a powerful ally of informal education. Over the years, we have learned a great deal about how to reach large, voluntary, home audiences with quality educational materials. It's not easy. Carefully planned uses of television are team efforts, requiring expertise not only in production but also in content, curriculum, educational research, evaluation, and local utilization. When a television show can succeed in reaching a voluntary audience with quality materials, the chances are good that those materials can be used in additional settings, such as schools, museums, science clubs, and other community organizations. The Children's Television Workshop is proud of what *3-2-1 CONTACT* is accomplishing, both on its own and in combination with other media and other organizations.

Helpful Hints

The following list of action-oriented ideas offers teachers a few suggestions for getting the most out of *3-2-1 CONTACT.* We recommend that teachers use their creativity and knowledge of their classes to develop original ideas and to tailor the suggestions to the unique needs of their students.

- Use the *3-2-1 CONTACT* Teacher's Guide. The Guide features daily show information, discussion questions, science vocabulary terms, elementary science bibliography, curriculum correlation chart, and reproducible activity pages. Guides are available for $5 each, or $2.50 each for 50 or more. Write to *3-2-1 CONTACT,* CTW, One Lincoln Plaza, New York, NY 10023.

- Consider alternative audiovisual formats for in-class use. *3-2-1 CONTACT* offers free three-year tape and erase rights for in-class audiovisual use. Additionally, Guidance Associates, in cooperation with the Children's Television Workshop, produces audiovisual programs based on *3-2-1 CONTACT* for schools. These programs, organized for classroom use, are available in sound filmstrip, videocassette, and 16-mm film formats. Call 1-800-931-1242, toll-free, for information.

- Encourage students to write scripts for *3-2-1 CONTACT* segments. This assignment places *3-2-1 CONTACT* in a multidisciplinary program, because scriptwriting requires students to apply their understanding of science content while sharpening their creative writing skills.

- Devote a day to careers in science and technology, as seen on *3-2-1 CONTACT.* Show and discuss segments that include role model scientists in their varied environments.

- Help students to organize a science fair revolving around *3-2-1 CONTACT* themes. Students watch assorted theme weeks, choose a topic, and design related science projects.

- Assist students while they create their own science video modeled after *3-2-1 CONTACT.*

- Recommend *3-2-1 CONTACT* for home viewing. Encourage students to summarize segments either verbally or in writing.

- Recommend *3-2-1 CONTACT* magazine for home reading. Write to *3-2-1 CONTACT* magazine, EBMC Square, P.O. Box 51177, Boulder, CO 80321-1177.

6 Behind the Scenes of Mr. Wizard

Don Herbert
Mr. Wizard Studio
Bell Canyon, California

Television viewers may consider Mr. Wizard to be a science teacher, but he's not faced with the same constraints as a typical teacher. He has only one "student," no set curriculum, no "principal," an adequate budget, more help, and "teaches" only one half hour a week. However, both have to impart some aspect of science to a young person. Perhaps some of the techniques we developed over the years for a television audience may help the teacher facing a classroom of students.

The question most often asked of me is: "Where do you get all your ideas?"

Primarily from published materials. Specialists in teaching science have described their techniques and collected the methods of others. I've found such volumes in bookstores, secondhand shops, and university stores. Two of the treasures in my library are *Experimental Science—Elementary Practical and Experimental Physics*, by George M. Hopkins, published in 1890, and

Magical Experiments or Science in Play, by Arthur Good, published in 1894. Modernized versions of many of the suggestions for science "experiments" described in the 1890s appear in today's books.

The Mr. Wizard library now numbers about 1,000 volumes, including eight sets of encyclopedias, science textbooks for elementary, junior, and senior high, and 20 file drawers of cataloged suggestions for science activities. From this collection we extract ideas for planning a "run down," the television version of a lesson plan. However, collecting demonstrations is only the beginning. They have to be selected and organized to make a "plot" that has a beginning, middle, and an end, and structured around a tried-and-true formula for drama:

- Catch attention.
- Arouse interest.
- Develop conflict.
- Resolve conflict.

Catching attention most often takes the form of a question, puzzle, problem, contradiction, or challenge for the child as he or she comes through the door at the opening of the show. The plot "thickens" and further interest is aroused as we work out the answer that leads to the next step in the plot.

It is important that the first challenge implies that finding the solution will be fun. Also, it should not be a test of knowledge or intelligence, but an invitation to explore. By posing an intriguing challenge, we can fairly easily catch attention and arouse interest.

Developing conflict is a greater problem. Certainly, there should be no conflict between the child and Mr. Wizard. What can we do that has some appearance of conflict without boy-meets-girl situations, murders to solve, or even the simplest kind of chase? The only motivation that seems appropriate is curiosity, an essential element in all dramatic situations in which the viewer wonders what is going to happen next. We set out to explore techniques for generating, sustaining, and satisfying curiosity.

To suggest possible plotting techniques, we have compiled a list of "trigger" words: puzzle, riddle, mystery, secret, challenge, test, investigation, quiz, dare, problem. Whenever we are stalled in coming up with a workable structure, we use these to cue our imagination. Most of the time, one will help generate a solution.

Once we select the series of demonstrations, we look for ways to make them more effective. This symbol focuses our thinking:

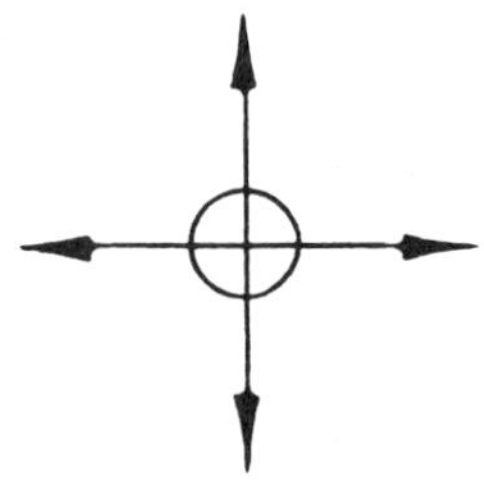

The circle represents the original idea. The vertical arrows suggest we improve it by making it higher, lower; taller, shorter; deeper, shallower; or heavier, lighter. The horizontal arrows suggest we consider making it faster, slower; bigger, smaller; or thicker, thinner. It is amazing how often such a simple device prompts us to make a "standard" demonstration more effective.

In the beginning, our restricted budget meant we had to use simple equipment we could find around the house or buy at a local store. Without realizing it, we were taking Dr. Karl Duncker's test for creativity week after week. Dr. Duncker, a German psychologist, challenged his students to attach three candles to a door without the candles touching the door. The students could use only boxes of birthday candles, matches, and thumbtacks on a table. Only a few figured out how to do it. Then he challenged another group, but put the candles, matches, and thumbtacks on the table out of their boxes. Many more students were able to solve the problem. The first group saw the boxes as containers, and that function became fixed in their minds. The second group could more easily visualize the boxes as platforms tacked to the door on which to set the candles. Dr. Duncker coined the term "functional fixation": the tendency to think of the properties of an object only in terms of the purpose for which the object was originally designed and to overlook other attributes that could make it useful in other ways.

A common example of function freedom is the use of soap. Its slipperiness is incidental to its primary purpose of getting things clean. That secondary property is the very thing needed to fix a sticky drawer. Other examples include using a screwdriver to open a paint can, holding a door open with a brick, making paste with flour, and cutting off the corner of an envelope to use it as a funnel.

We first define the specific properties we need and then begin looking for everyday items that have those properties. Function freedom guides us as we walk the aisles of supermarkets, hardware stores, drugstores, and hobby shops looking for "scientific" equipment. With functional freedom a clothespin becomes an electrical switch, a sparkler an example of the randomness of radioactivity, a marshmallow an illustration of a solid colloid filled with a gas, and gumdrops atoms in a model.

Using common items has another advantage. As an analogy for 1½ volts from a battery and the 120 volts from a wall socket, we use a mousetrap and a rattrap. When both are set and the child is ready to spring them with a soda straw, the difference in potential energy between them is apparent. The instantly recognizable rattraps and mousetraps and a soda straw do not have to be "explained." The child can anticipate the results and see and hear the difference in "voltages" as the traps are sprung. Using an everyday item in a novel way to explain a scientific phenomenon has become part of Mr. Wizard's identity.

Later, we discovered that the techniques we had developed for planning a science television show had been validated by educational researchers for use in the classroom. Dr. John Cunningham, of Keene State College, New Hampshire, in *School Science and Mathematics,* December 1966, described scores of investigations into what made effective teaching and concluded:

> The upshot of all these currents of thought converging from different starting points is that motivating disturbances can come from conflict and that conflict often arises as a result of curiosity . . . the curious person not only sees and hears but *looks for* and *listens for.*

He goes on to quote Dr. D. E. Berlyne, of the University of Toronto:

> The element of enduring value in the new teaching techniques, including the discovery method and programmed instruction, the secret of whatever success can be claimed for them, the real germ of a pedagogical revolution, may well turn out to be not in independent discovery, but in the attempt to pinpoint and harness the sources of epistemic curiosity.

Dr. Cunningham concludes that "many of the new science and mathematics curricula work largely by manipulating conceptual conflict in the form of surprise, doubt, incongruity, perplexity, and confusion."

Mr. Wizard and science teachers do have much in common after all.

Even Television Experiments Aren't Foolproof

A question that's often asked, especially by reporters, is the following: "In view of the live nature of the show, did anything ever go wrong?"

Often enough, in spite of our careful planning and rehearsal, it was obvious from the way the child and I talked to each other and how action unfolded that the program was adlibbed and "live." No doubt the audience recognized and responded to this atmosphere of uncertainty. When something unintended happened, we did whatever was necessary to remedy it, and continued with our "experiment" just as any two people would in "real life."

However, failure can be the springboard for success. Our failure with rockets is a good example. The climax of a show on rocketry was, of course, firing a rocket. We rigged a hobby store version to fly across the set on a wire. The fuse was a tricky affair that sputtered and went out. The climax was a dead rocket that flew nowhere. We adlibbed our way to the end of the show, vowing to try again next week.

On the following week's show, we started with a "foolproof" rocket made of pipe that had a dynamite-type fuse to ignite it. The fuse burned and ignited the fuel, but the pipe got hung up on a wire and stayed on the launchpad. We went on with the rest of the show.

In preparation for our third attempt, we tried the functional freedom approach that had worked for us in the past. What common item was made of metal to withstand the burning of the fuel, had a large enough opening through which to load the fuel, yet could be closed tightly enough to contain gas under pressure? It also had to have a small hole through which the gas could escape to create the thrust to propel the rocket. The method of igniting it also had to be foolproof. The properties we defined fit a common hardware store item perfectly. Our rocket was a small oil can. We took off the spout to load it, screwed it back on and tightened it with a pair of pliers, tethered the can to a pole, and ignited it with a small blowtorch aimed at the bottom. At the opening of next week's show, it launched successfully. We've repeated the oil-can rocket experiment many times, and it has always worked.

The most potentially serious mishap occurred during the well-known demonstration in which a burning candle is covered by a bottle with the mouth of the bottle underwater. As the candle flame goes out, the water flows up into the bottle, supposedly because the flame used some of the oxygen in the air in the bottle, producing a partial vacuum. Atmospheric pressure forces the water up into the bottle to take the place of the missing oxygen. This explanation is incorrect. Actually, the flame heats the air in the bottle, expanding it enough so that some of it bubbles out of the mouth. Most of the time, this loss of air goes unnoticed. We figured out a way to have the child deduce the correct explanation by lighting the candle after it was inside the bottle.

On the show we did the experiment in the standard way and outlined the usual explanation. We then produced our new version, in which we had placed a small coil next to the wick of the candle. We could cover the candle with the bottle, and with a tube, suck out enough of the air from the bottle to have water rise partway into it. We sent an electric current to the coil, which heated it red hot, and ignited the candle. In rehearsal, the heat of the flame expanded the air, forcing the water level down almost to the mouth of the upside-down bottle. When the flame went out, the water rose to its previous level, showing that the standard explanation was incorrect.

Still in rehearsal we pointed out that the volume of the gases produced by the flame was almost the same as the volume of gases before the candle was lighted. The child gave the correct explanation: The candle flame heated the air and some of it escaped from the bottle. We were pleased that we had designed a new demonstration that neatly proved the correct explanation. Little did we realize disaster was about to strike.

As we set up for the actual telecast, we removed the bottle from the candle without realizing that the wick got wet. On the air, when we turned on the current to the coil, it got hotter and hotter, but the wick did not ignite. Instead, the air in the bottle became filled with vaporized paraffin. When the coil finally got hot enough, it ignited the mixture, blew the bottle out of the close-up, and sprayed water over the child and me. The bottle came down out of the lights and clanged to the floor. As soon as the crew realized that neither the child nor I was hurt, they roared with laughter. From long training in doing a live show, I wiped the two of us off, explained that I got the wick wet, and described briefly what should have happened. In the classroom, the teacher could have put a new candle in position and tried again. With other demonstrations scheduled and timed, we could only go on to the next one.

The accident happened in the final days of Mr. Wizard when the program was recorded on tape for later rebroadcast. It is the only time the show was ever edited. They cut only the laughter of the crew, and the program went on the network with the accident intact.

Television Science Can Have Lasting Influence

While the show was on the air we had little real contact with the audience and had no idea of the show's impact. Now, many years later, many adults have told me that as child viewers they were influenced by the show. Some have said it taught them to think or that it kindled an interest in science. Some have even gone so far as to say it was responsible for their careers in science or technology. How could a half-hour television program in black and white seen on small screens have had such an influence?

One reason is that the program was on long enough for us to be able to experiment and refine our techniques. The series was also on long enough for viewers to respond to the informal style and the scientific content and to find the show was worth a half hour of their time.

Another reason for the show's impact is probably because we included activities that children could repeat at home. This meant that they had to think about what they were doing, which made them better able to recall the principle involved and perhaps look forward to the suggestions on the next show. Repeating some of the "tricks" also meant the children were not merely inactive spectators, but could actually enjoy doing science. Learning implies there's been an internal change in the learner. As children tried to repeat what they'd seen us do on television, they were, in effect, teaching themselves.

If a half-hour show with contact only through a television screen could have such influence, think of the tremendous opportunity that a science teacher has with more time and that very important element of close personal communication.

The original show aired on NBC from 1951 through the middle of 1965. Now, some 23 years later, it is back in *Mr. Wizard's World* on *Nickelodeon*, the all-children's programming cable channel. We applied everything we learned on the old shows to generate and satisfy the curiosity of a new generation of children and their parents, who were the children who watched the original so many years ago.

The most important ingredient, a child having fun experimenting with things scientific and technical, is now incorporated into a magazine format more in keeping with the television of today. Interspersed with our usual activities are "Quick Quizzes" (short puzzlers), "Snapshots" (intriguing pictures), "Safaris," and "Science Frontiers" (scientific and technological explorations on film and tape). The series is in color and has more variety and a faster pace. Titles and music introduce the show and each segment, and a stock company of children of various ages stamp each segment with their personalities. There are three indoor playing areas, and many of the sequences are taped on location. The uncertainty of the adlib style is still there, but now it's on tape and can be edited for better control and pacing.

Television Science Is for Adults As Well

We're contributing to continuing education for adults in another kind of television program: newscasts. In 1980, the National Science Foundation (NSF) challenged us to find a way to bring science and technology to the vast audience that watches commercial television and is unlikely to tune into a typical science program. My wife,

Norma (who is in charge of marketing our activities), and I, after researching the problem, realized that the only programming that could present science on a regular basis in peak viewing periods was a newscast. With our first grant from the NSF, we followed the advice of news directors and produced a series of 90-second reports.

The series is titled *How About . . .*, because our goal is to have the viewer say *"How about that!"* The series started on the air in the spring of 1980. To date, 621 reports have been produced. Each year for the past nine years an average of 140 commercial television stations have aired the series. For the last six years, it also reached an international audience via the Armed Forces Radio and Television Network. The reports cover a wide range of subjects by highlighting the latest developments in science and technology at universities and other research-oriented organizations across the country. The project is a good example of continuing informal education made possible by the cooperation of government and industry: the NSF, an agency of the federal government, and General Motors, one of the leading corporations in the private sector.

We are now exploring how the series can be made available to teachers. On videocassette many of the reports can bring students up-to-date on developments in some of the disciplines covered in the curriculum. With appropriate supplementary materials, they could also be stimuli for discussion.

While the subject matter in the *How About . . .* series is different from that of the *Mr. Wizard's World* programs, some of the same techniques for catching attention, arousing interest, and developing and resolving conflict in the form of curiosity are again helping us bring a better understanding and appreciation of science and technology to millions of television viewers.

References

Cunningham, John. (December, 1966). On curiosity and science education. *School Science and Mathematics.*

Good, Arthur. (1894). *Magical experiments or science in play.* Philadelphia, PA: David McKay.

Hopkins, George M. (1890). *Experimental science—elementary practical and experimental physics.* New York, NY: Munn & Co.

7 Science Writing in Its Age of Yellow Journalism

Jon Franklin
University of Maryland
College Park, Maryland

Two classic images of the scientist have long vied for dominance in the American mind. In one guise, the scientist is the absentminded professor, the bumbling, over-educated fool who has to be reminded to put on his galoshes. In his alter ego, he is Dr. Frankenstein, the mad scientist, the amoral if not immoral creator of a monster.

At a certain level, the public believes that scientists are people who mumble in languages they don't understand, tap strange forces, think in terms of cohorts of rats and tons of cow pancreases reduced to test tubes full of extract—and who produce, as if by magic, the thermonuclear bomb.

People in science are quick to jump to the conclusion that the media shapes these stereotypes. A University of Pennsylvania group found that people who watch a lot of television think of science as a threatening enterprise. Television scientists frequently are portrayed as the bad guys. Yet the media capitalizes on attitudes that already exist; they do not create the negative view of science, but make a bad situation worse by reinforcing it.

The media's willingness to capitalize on anti-scientific feelings spreads from prime-time television to the news hour, and eventually into respectable newspapers. Newspaper reporters and editors feature science stories on "scare topics" such as the Dalkon Shield, Agent Orange and dioxin, hazardous waste, acid rain, nuclear winter, and genetic engineering, often at the expense of covering important advances in science. They didn't cover the neurotransmitter story until long after Sol Snyder and Candace Pert made their watershed discovery. And what about the genetic engineering of plants? Materials science? Agronomy? We can all think of a long list of scientific trends and discoveries that the public should know about and the press should write about extensively.

Part of this problem is certainly due to a lack of education among commentators, reporters, and editors. Media folks are as a rule ignorant of science—even more so than the public at large. They're not dumb—don't get me wrong. But as the phrase "the fourth estate" implies, the tradition of journalism tends to ally the journalist with the politician and the social administrator. The politician and the reporter need each other, they know they need each other, and they make a big effort to understand each other.

No such tradition exists between scientists and journalists. Most reporters and editors do the vast majority of their college course work, for instance, in political science. The average journalist takes one required course, say, in chemistry—and finds it so difficult and obscure that he or she is terrified of flunking. The journalist survives mostly by memorizing things, and learns the lesson that science is awful, that it can't be understood by normal people, and that at the same time, it's very important.

And we're back, once again, to the idea that scientists run things and the average Joe or Jane, in this case the newspaper reporter, is incapable of understanding science, much less of influencing the outcome. In my experience, that paranoia is rampant in the American newsroom: Science is seen as dangerously out of control. Many science writers see it that way, and almost all environmental writers (who outnumber science writers and usually get much better play) believe it to be so. Editors, by and large, are critical and suspicious of science and of all its products. And most science stories are covered by general reporters, *not* by science writers.

Mutual Fear and Hostility

So the public fears science and the media not only plays into that fear but shares it. And the average scientist is often as ignorant, frightened, and judgmental about the media as the media is about science. My favorite example of this predicament goes back to the 1950s, when the first pulsar was discovered. The United States had built the world's largest radio telescope at Arecibo in Puerto Rico. It was too big to aim, so it was nestled in a natural crater so that it pointed straight up. Each day, as the earth turned, it swept across the heavens. Fairly quickly the scientists discovered that each time the dish swept across one specific section of sky, it picked up a series of sharp, short bursts of radio energy. Something up there was beeping at them, very fast, beep-beep-beep-beep-beep.

We know now that it was a pulsar, the cinder of an exploded star. It was spinning incredibly fast and giving off a radio beacon. With each revolution, a beep was heard. But the Arecibo scientists knew only that something up there was beeping at them. And since they had never come across anything like that before, it *had* to be important. They knew they had to announce their discovery to the world, but weren't too sure how to go about it. Eventually they decided to find the largest newspaper in the United States and call an editor there. The best-read newspaper in the United States turned out to be a publication printed in Lantana,

Florida, just a few hours' flight to the north. A while later, in Lantana, an editor picked up a telephone. Yes, he *was* interested, and would send one of his best reporters on the next plane south.

The astronomers at Arecibo let the reporter listen to the tape-recorded beeps and gave him a lesson in radio astronomy. Finally, he asked, "So you don't really know for sure what's doing the beeping?" No, they replied. All they really knew was the beeping. The reporter said, "Then it could be anything, right? It could even be intelligent life trying to start a conversation."

The scientists laughed, pleased that the fellow was beginning to grasp the basic vagueness of science. They said, "Sure, we suppose it could be anything." They started talking about the wonders of radio astronomy again, but the reporter was suddenly in a hurry to get back to his newspaper.

And so, the most important astronomical discovery of the decade appeared on the front page of *The National Enquirer.* The banner headline announced, in 72-point type, something like "Aliens Contact Earth."

One of the scientists involved told me this story. He'd learned his lesson—he was never again going to talk to a reporter. It was 10 years before he cooled down enough to understand that the fiasco was his own fault for not making distinctions between different newspapers.

We have this deadly combination of ignorance and hostility on all fronts. We have scientists who won't cooperate, reporters who don't know what they're writing about, editors who don't care, and readers with so many misconceptions and preconceptions that they wouldn't know an important story if it bit them. This is a difficult truth, and during the first part of my career as a science writer I stubbornly resisted it. I wanted to be a SERIOUS journalist. I wrote well-researched, carefully crafted hard news stories about significant scientific advances. These stories usually appeared on page D-43, cut by half and draped gracelessly around a tire ad. Nobody noticed them.

The reason? Such technical science stories are very precise, but they have one major drawback. They are written with the assumption that the reader has a context for the information. But the story actually will make sense only to a small subset of people who have the background to understand it. Consequently, many important discoveries remain unknown to all outside the writer's immediate professional circle. For instance, the revolutionary breakthroughs in neurochemistry remain even today largely incomprehensible to most people.

Furthermore, such hard news stories don't relate to the reader and his or her life. The public doesn't want to read about "serious" science, at least the way I was presenting it. The undeniable fact was that I had chosen to be a science writer in the middle of science writing's own age of yellow journalism, and I could either fight it or join it. I chose to join it.

The Age of Yellow Journalism

The age of yellow journalism is a period of history that lasted about a generation, from the 1800s to the early 1900s, and was characterized by emotionalism. About the only examples left are tabloids—*The National Enquirer, Weekly World News*—that you see in the supermarket checkout lines.

Yellow journalism didn't develop in a vacuum. In this period the majority of Americans became truly literate. Before the 1800s, two intellectual classes existed in this country: the very well-educated and the very illiterate. Periodicals were directed at the relatively small, upper-class audience. What about the ones who didn't read those papers—the illiterate majority? First, keep in mind that by "illiterate" I don't really mean that they couldn't read and write. Literacy, for civic purposes, requires more.

Most Americans had only the vaguest idea how their government worked. They understood little of the historical context of their institutions, let alone the flow of power within them. As to theories of government, they knew America was the best in the world. To be a good citizen in those days was enough. Then society began to change. In 1870 seven million children went to public schools; during the next 30 years, that number doubled.

A peculiar thing happens with a little bit of learning. People want to learn more. Reading makes them want to read more. Suddenly, an incredible hunger developed for the written word, and a market for a true mass media was born. This was made possible by the invention, about the same time, of the high-speed printing press.

But—and this is very important—the demand wasn't for just any written material. The vaguely literate, wanting to understand but without much knowledge, read what grabbed them. What grabbed them was stuff they could relate to, that avoided the abstract. What was not abstract? An axe murder. A theater fire. A juicy scandal, something with a hint of sex in the wings. A market existed not only for the scandal story, but especially the scandal story that involved the perfidy of some powerful figure. The public wanted what today we call the "gotcha" story—the story that exposes corruption, malfeasance, lying, cheating, and general public-be-damned attitudes on the part of the establishment. The writers of these stories were called muckrakers. Many of these stories were merely sensationalistic, focusing on sex and violence, and some stories turned the public against politicians and public officials who were trying to do a good job. But at the same time, crusading reporters destroyed the Tweed empire and uncovered inhuman working conditions, and the government generally became more accountable.

Another kind of story grabbed the public, too—the drama story, usually a legitimate story of legitimate value with an emotional base. Balloon races made good copy. So did polar exploration and mountain climbing. People read exploration stories, for example, because they had emotional appeal. But along with the excitement of reading about explorers, readers (whether they knew it or not) were also getting an introduction to geology, politics, and world cultures.

Even more important was the part that yellow journalism played in the steadily rising literacy rate. Marginally literate people may appear to gain nothing specific from reading a good, bloody story about a ball pein hammer murder, yet they do gain something just by reading. They become familiar with the language, and if they misunderstand the institutions that are involved, at least they learn that such institutions exist.

The age of yellow journalism represents the adolescence of the American press. As the audience slowly became more sophisticated, it demanded more. A reporter doesn't dare know less than his or her readers. Newspapers also became more sophisticated—large, popular, and financially successful. This age of experimentation culminated in the tradition of journalism you see today in papers like *The New York Times*, *The Baltimore Sun*, and *The Washington Post*. It was an age of floundering around and learning by experience, of literacy-building, and of incorporating ever larger percentages of the population into the system.

That, precisely, is where science writing stands today.

Science Writing That Entertains

My theory was that the public really does want to know about science. They know technology is our master today, and science is the master of technology. If you're going to get ahead, you've got to learn to deal with computers. If you're going to get a job, you've got to know science. The public is curious about science, but they want it in some context they can understand. And the scientific story, whatever it is, must entertain.

As I considered the problem of communicating across the great chasm between science and society, of talking real science to police officers and bank tellers and English professors, I realized that I would need to tap a new level of power. I would need to make a quantum jump into something called "emotion."

Essays. I began to use the essay form, a type of

writing that helps readers identify with the subject and appeals to a basic emotion—curiosity. A good essay leaves readers with the satisfying feeling that they know something they didn't know before, perhaps something they believed they were incapable of understanding. Done correctly, essays can elicit from the reader a certain pantheistic, quasi-religious awe. In the process, the reader's mind is open to legitimate information and perspectives. Essays tap the reader's sense of wonder and entice him or her into reading about the mystery of black holes or the intricate mechanism of the human body. The idea is to have readers think, "Gee whiz, look at the universe! Look at how marvelous it is, how well-ordered, how understandable!" Instead of leaving readers with the perception that science is a cold and inhuman process, and that *real* people can't understand it, essays allow readers to believe scientific research is a fascinating endeavor. The reader can participate, the same way he or she can participate in music by developing a taste for Beethoven.

So, besides being entertaining, essays provide an easy entry into science for the nonscientific. The reader puts down the piece with the distinct feeling that despite a bad experience in college chemistry, he or she can in fact understand science. The essay form has the power to reach across the great chasm, touch the nonscientist, and change minds.

The drawback is that it can change only the minds of those people who will read essays that treat science with respect. It will not work in minds that are already highly paranoid and suspicious about science, because they will see the work as propaganda. For example, people who think medical scientists are concerned primarily with making money don't become fans of Lewis Thomas, author of *Lives of a Cell* and other books of essays on the mechanisms of science.

Stories. I would not be satisfied until I won the attention of the majority of folks who really don't care about science one way or the other, except to fear it. To do this, I began to look for stories in science that could be told in classic short-story form.

Dramatic form always entails more than an idea. It features a character or characters. In the course of the story the character confronts a problem. As the story continues, the character struggles to face and overcome the problem. The power in the dramatic form of writing is tapped through the phenomena of *suspension of disbelief*, which induces *experience*. Readers forget they're sitting in an easy chair, or a bus, or wherever. Readers *become* the character, sharing the character's experiences, world, excitement, tension, sorrow, and ultimate fate. My character was usually a scientist, often a doctor-scientist, and the conflict was usually with the unknown—humans against the rest of the universe.

Experience, as we all know, is the ultimate teacher, the ultimate giver not of information but of knowledge, the ultimate shaper of pattern and world view. Therefore drama, unlike any other art form, can actually alter the reader's view of the world. It has the potential to heal the cultural wound, the science/society rift.

In science writing's own age of yellow journalism, this technique proves to be powerful. The stories have high entertainment value. They attract and hold the reader for their own sake. The plot is the important element, not the science, but if the reader learns some science along the way, that's great. But it isn't why the reader reads the stories. I try to keep "science" out of the title.

When I present science in such a human context, readers not only understand what they read but *remember* it. In these stories, science is revealed as a human thing, pursued by human beings not unlike the readers. I get a lot of positive feedback; people say they like the people I write about, and the stories make them much more in tune with science and much less afraid of it. Some readers say that I've even changed their attitudes about science.

These ideas aren't exclusively mine. In recent

years more and more science writers are coming to share them, and science stories are rapidly becoming the standard in science journalism.

Science Literature

You've probably heard that Truman Capote invented what he called the "nonfiction novel." Yet the first true example of the form that I've been able to find is *Hiroshima*, a book written in the 1940s by John Hersey, a crossover novelist and journalist. The book focused on the stories of people who were in Hiroshima when the atomic bomb was dropped. It is first-class literature, in form as well as function, and its aim is to make a technological product understandable to the reader. The effect is to put the reader there, to constitute an *experience.*

Another pioneering science writer was Loren Eiseley, a physical anthropologist who wrote from the University of Pennsylvania. Although he's best known for his classic book about evolution, *The Immense Journey*, he wrote dozens of other books, all about science. Many of them are studied today in English literature classes.

In recent years the market for literary works of science (dramatic nonfiction and essays) has been growing. One of the most interesting recent success stories involves Tom Wolfe. Wolfe, after laboring with a limited audience for years, finally moved into science—specifically, space science. His book *The Right Stuff* put him on the bestseller list and made him an international celebrity.

The art of science writing is also attracting a substantial number of physicians and scientists. Lewis Thomas, a physician and head of Sloan-Kettering, was the first one to put a literary, essay-form book (*Lives of a Cell*) on the bestseller list. Through him, Americans became aware for the first time of cellular biology. More recently, Oliver Sacks, a neurologist, rang the commercial bell with *The Man Who Mistook His Wife for a Hat*, a book relating strange tales from the neurological frontier.

A whole generation of younger writers are using science and scientific subjects to develop more modern, commercially viable literary styles. And I'm talking about very state-of-the-art subjects. One such writer is Douglas Hofstadter, who in 1964 won a Pulitzer for his book *Anti-Intellectualism in American Life.* In 1980 he won again for a stream-of-consciousness nonfiction book entitled *Gödel, Escher, Bach, an Eternal Golden Braid.*

Many science books have won the Pulitzer prize in the General Nonfiction category over the last decade. In 1977, the judges picked *Beautiful Swimmers*, by naturalist William Warner, who used the biology of the blue channel crab as a literary carrier vehicle to probe the culture of Maryland's Eastern Shore. The following year, Carl Sagan won with *The Dragons of Eden.* In 1979, the winner was Edward O. Wilson, for *On Human Nature.* In 1980, as I've just mentioned, Hofstadter won for *Gödel, Escher, Bach, an Eternal Golden Braid.* In 1982, we had Tracy Kidder's *The Soul of a New Machine*, and in 1984, Paul Starr's *Social Transformation of Modern Medicine.*

* * * * *

From the public, especially from today's students, come tomorrow's journalists, editors, politicians, and commentators. It's our job to provide the public with science stories and essays that will be read, understood, and remembered—that entertain as well as educate. These stories and essays assuage the public's fear of science by letting them into the laboratories to share the drama of science with scientists.

8

Dusk in the Robot Museums: The Rebirth of Imagination

Ray Bradbury
Los Angeles, California

For some 10 years now, I have been writing a long narrative poem about a small child in the future who runs into an audio-animatronic museum, veers away from the right portico marked Rome, passes a door marked Alexandria, and enters across a sill where a sign lettered Greece points in across a meadow.

The child runs over the artificial grass and comes upon Plato, Socrates, and perhaps Euripides seated at high noon under an olive tree sipping wine and eating bread and honey and speaking truths.

The child hesitates and then addresses Plato: "How goes it with the Republic?"

"Sit down," says Plato, "and I'll tell you."

The child sits. Plato tells. Socrates steps in from time to time. Euripides does a scene from one of his plays.

Along the way, the child might well ask a question that hovered in all of our minds during the past few decades:

"How come the United States, the country of Ideas on the March, for so long neglected fantasy and science fiction? Why is it that only during the past 30 years attention is being paid?"

Another question might well be:

"Who is responsible for the change?

"Who has taught the teachers and the librarians to pull up their socks, sit straight, and take notice?"

Since I am neither dead nor a robot, and Plato-as-audio-animatronic lecturer might not be programmed to respond, let me answer as best I can.

The answer is: The students. The young people. The children.

They have led the revolution.

The children have become the teachers. Before our time, knowledge came down from the top of the pyramid to the broad base where the students survived as best they could. The gods spoke and the children listened.

But, lo! Gravity reverses itself. The massive pyramid turns like a melting iceberg, until the boys and girls are on top. The base of the pyramid now teaches.

How did it happen? After all, back in the twenties and thirties, there were few science-fiction books in the curricula of schools. There were few in the libraries. Responsible publishers only dared to publish books that could be designated as speculative fiction.

If you went into the average library as you motored across America in 1932, 1945, or 1953, you would have found:

No Edgar Rice Burroughs.

No L. Frank Baum and no Oz.

In 1958 or 1962 you would have found no Asimov, no Heinlein, no Van Vogt, and, er, no Bradbury.

Here and there, perhaps one book or two by the above. For the rest: a desert.

What were the reasons for this desert?

Among librarians and teachers there was then, and there still somewhat dimly persists, an idea, a notion, a concept that only Fact should be eaten with your Wheaties. Fantasy? That's for the Fire Birds. Fantasy, even when it takes science-fictional forms, which it often does, is dangerous. It is escapist. It is daydreaming. It has nothing to do with the world and the world's problems.

So said the snobs who did not know themselves as snobs.

So the shelves lay empty, the books untouched in publishers' bins, the subject untaught.

Comes the Evolution. The survival of that species called Child. The children, dying of starvation, hungry for ideas that lay all about in this fabulous land, locked into machines and architecture, struck out on their own. What did they do?

They walked into classrooms in Waukesha and Peoria and Neepawa and Cheyenne and Moose Jaw and Redwood City and placed a gentle bomb on teacher's desk. Instead of an apple it was Asimov.

"What's that?" the teacher asked, suspiciously.

"Try it," said the students. "Read the first page. If you don't like it, stop."

And the clever students turned and went away.

The teachers (and the librarians, later) put off reading, kept the book around the house for a few weeks and then, late one night, tried the first paragraph.

And the bomb exploded.

They not only read the first but the second paragraph, the second and third pages, the fourth and fifth chapters.

"My God!" they cried, almost in unison, "these damned books are about something!"

"Good Lord!" they cried, reading a second book, "there are Ideas here!"

"Holy Smoke!" they babbled, on their way through Clarke, heading into Heinlein, emerging from Sturgeon, "these books are—ugly word—relevant!"

"Yes!" shouted the chorus of kids starving in the yard. "Oh, my, yes!"

And the teachers began to teach, and discovered an amazing thing:

Students who had never wanted to read before suddenly were galvanized, pulled up their socks, and began to read and quote Ursula K. Le Guin. Kids who had never read so much as one pirate's obituary in their lives were suddenly turning pages with their tongues, ravening for more.

Librarians were stunned to find that science-fiction books were not only being borrowed in the tens of thousands, but stolen and never returned!

"Where have we been?," the librarians and the teachers asked each other, as the Prince kissed them awake. "What's in these books that makes them as irresistible as Cracker Jacks?"

The History of Ideas.

The children wouldn't have said it in so many words. They only sensed it and read it and loved it. The kids sensed, if they could not speak it, that the first science-fiction writers were cavemen and women who were trying to figure out the first sciences—which were what? How to capture fire. What to do about that lout of a mammoth hanging around outside the cave. How to play dentist to the sabre-tooth tiger and turn it into a housecat.

Pondering those problems and possible sciences, the first cavemen and women drew science-fiction dreams on the cave walls. Scribbles in soot blueprinting possible strategies. Illustrations of mammoths, tigers, fires: How to solve? How to turn science-fiction (problem solving) into science-fact (problem solved).

Some few brave ones ran out of the cave to be stomped by the mammoth, toothed by the tiger, scorched by the bestial fire that lived on trees and devoured wood. A few finally returned to draw on the walls the triumph of the mammoth knocked like a hairy cathedral to earth, the tiger toothless, and the fire tamed and brought within the cave to light their nightmares and warm their souls.

The children sensed, if they could not speak, that the entire history of humankind is problem solving, or science fiction swallowing ideas, digesting them, and excreting formulas for survival. You can't have one without the other. No fantasy, no reality. No studies concerning loss, no gain. No imagination, no will. No impossible dreams: no possible solutions.

The children sensed, if they could not say, that fantasy, and its robot child science fiction, is not escape at all, but a circling round of reality to enchant it and make it behave. What is an airplane, after all, but a circling of reality, an approach to gravity that says: Look, with my magic machine, I defy you. Gravity be gone. Distance, stand aside. Time, stand still, or reverse, as I finally outrace the sun around the world in, by God! look! plane/jet/rocket—80 minutes!

The children guessed, if they did not whisper it, that all science fiction is an attempt to solve problems by pretending to look the other way.

In another place I have described this process as Perseus confronted by Medusa. Gazing at Medusa's image in his bronze shield, pretending to look one way, Perseus reaches back over his shoulder and severs Medusa's head. So science fiction pretends at futures in order to cure sick dogs lying in today's road. Indirection is everything. Metaphor is the medicine.

Children love cataphracts, though do not name them thusly. A cataphract is a special Persian on a specially bred horse: The combination threw the Roman legions back some long while ago. Problem solving. Problem: massive roman armies on foot. Science fiction dreams: cataphract/human-on-horseback. Romans dispersed. Problem solved. Science fiction becomes scientific fact.

Problem: botulism. Science fiction dreams: to someday produce a container that would preserve food, prevent death. Science-fictional dreamers: Napoleon and his technicians. Dream becomes fact: the invention of the Tin Can. Outcome: millions alive today who would have otherwise writhed and died.

So, it seems, we are all science-fictional children dreaming ourselves into new ways of survival. We are the reliquaries of all time. Instead of putting saints' bones by in crystal and gold jars, to be touched by the faithful in the following centuries, we put by voices and faces, dreams and impossible dreams on tape, on records, in books, on television, in films. Humans are problem solvers only because they are idea keepers. Only by finding technological ways to save time, keep time, learn from time, and grow into solutions, have we survived into and through this age toward even better ones. Are we polluted? We can unpollute ourselves. Are we crowded? We can de-mob ourselves. Are we alone? Are we sick? The hospitals of the world are better places since television came to visit, hold hands, take away half the curse of illness and isolation.

Do we want the stars? We can have them. Can we borrow cups of fire from the sun? We can and must light the world.

Everywhere we look: problems. Everywhere we further deeply look: solutions. The children of humans, the children of time, how can they not be fascinated with these challenges? Thus: science fiction and its recent history.

And because the children of the Space Age wanted their fictional dreams sketched and painted in illustrative terms, the ancient art of storytelling, as acted out by your caveman or woman, was reinvented as yet the second giant pyramid turned end for end, and education ran from the base into the apex, and the old order was reversed.

Hence your Revolution.

Hence, by osmosis, the Industrial Revolution and the Electronic and Space Ages have finally seeped into the blood, bone, marrow, heart, flesh, and mind of the young, who as teachers teach us what we should have known all along.

I hope we will not get too serious here, for seriousness is the Red Death if we let it move too freely among us. Its freedom is our prison and our defeat and death. A good idea should worry us like a dog. We should not, in turn, worry it into the grave, smother it with intellect, pontificate it into snoozing, kill it with the death of a thousand analytical slices.

All of us would, or should, like to remain childlike and not childish in our 20-20 vision, borrowing such telescopes, rockets, or magic carpets as may be needed to hurry us along to miracles of physics as well as dream.

The Revolution continues. And there are more revolutions to come. And there will always be problems. Thank god for that.

Carbon Filament

Museums and Zoos

9 The Zoo as Noah's Ark

Annette Berkovits
Bronx Zoo
Bronx, New York

That people naturally think of zoos as sources of scientific information is perhaps best illustrated by excerpts from the scores of letters that arrive in the Bronx Zoo's mailbag. A seven-year-old who heard that armadillos ate termites wrote to check if the zoo could sell him an armadillo for his grandmother's birthday. He wanted to help her with a termite problem in the basement. Another youngster wanted to know where he could find the biological clock. A girl who discovered the concept of deciduous trees wanted to know if one could properly refer to "deciduous antlers." An aficionado of primates, aware of their social nature, asked whether a lone ape on exhibit could die of "melancholy." A teacher wanted to know how to take care of the egg she claimed her gerbil had laid.

Amusing as these inquiries are, they reflect the pervasive scientific illiteracy of the public. On the positive side, they also reveal how animals can evoke genuine scientific investigation, curiosity, and interest. Letters such as this one from a second-grader in California, who shared with us his letter to President Reagan, are extremely rare:

> Dear Mr. President, I am concerned about people who are killing endangered animals. I want you to help the animals by telling workers not to build houses where animals live. Animals have no place to go. And also, please stop people from shooting the animals. I want my children to see the animals. I am begging you please!

Thus, along with their former role of interpreting the mysteries of animal behavior and biology to the public, zoos are now being challenged to elicit an affective response from visitors. The new obligation to involve not only the visitors' minds, but their hearts as well, is rooted in the hope that visitors will be moved toward positive actions on behalf of the environment after their zoo visits. No longer is it sufficient for modern zoogoers to passively observe and admire. Zoos are increasingly expecting them to behave in certain ways. Gawking must be replaced by giving, admiration must translate into action, enjoyment must be coupled with understanding.

Ecological changes, particularly those since the Industrial Revolution, have influenced the role of zoos. None are more serious and none have had as significant an impact on the course of modern zoological institutions than the dramatic twentieth-century changes in the environment. Dr. William G. Conway (1984–85), General Director of the New York Zoological Society, which operates the Bronx Zoo, speaks eloquently about these changes:

> Among the largest issues facing science and sociology is the loss of wild animals and plants which is taking place throughout the world at a constantly accelerating rate. The next 50 years will see the extinction of a significant percentage of all remaining species. No matter what efforts are made to control the human population, the continuing loss of habitat is inevitable, if only because the humans who will occasion much of it have already been born.

Compelling zoo exhibitions such as Jungle World at the Bronx Zoo enhance visitors' opportunities for learning not only facts, but values. Modern zoos have suddenly been propelled into an intensive search for better and more evocative methods of exhibition and interpretation. With dramatic decline of wild places and wild species around the world, the work of modern zoos has taken on aspects of an emergency rescue operation. This is a new role for zoos, which until recently have served mainly as repositories and living libraries. Perhaps this newest era for zoos can aptly be named by a metaphor ahead of its time: the zoo as Noah's Ark.

The Zoo Educator's Dilemma: Quality Experiences for the Masses

There has been no better time in human history than the present for zoos to work with educational systems to affect the course of future existence of both animals and people.

Sounds lofty? Impossible? Perhaps not when you consider that more people attend today's zoos than all sports events combined. The combined annual zoo attendance in U.S. zoos is estimated to be 112 million. Worldwide, zoos have the same magnetic and universal appeal. During a recent tour of Chinese zoos, I learned that the annual attendance of the Beijing (Peking) Zoo is nearly 12 million. Numbers pale when one can experience firsthand the palpable excitement of zoo visitors as I did during an early morning visit to the Canton Zoo, where thousands of children were jockeying for positions in front of their favorite exhibits.

I encountered similar scenes in zoos from Warsaw to Madrid, from Belize to Tokyo, and closer to home in Topeka, Columbus, San Francisco, and Philadelphia. Everywhere, zoo visitors exhibit the universal behaviors associated with pleasurable learning.

If one were able to calculate precisely the worldwide zoo attendance figures in a single year, the potential and opportunity that zoos have for affecting people's attitudes toward wildlife conservation would be obvious and staggering. These numbers represent both the promise and problem for informal learning in zoos.

For most zoo visitors, viewing an exhibit begins with the ubiquitous question "What is it?" followed by an average 60 seconds of looking, and ends with "What's next?" or "Where is the bathroom?" What can zoo educators do to enhance the attention span of zoo visitors? First and foremost, educators must present animals and their fascinating behaviors in an interesting and relevant context. Everyone can recall at least one zoo label that showed a faded map of some obscure land mass along with an unpronounceable Latin name for the animal. "Au courant" zoo educators do not consider this to be proper interpretive technique.

How can we best educate in an informal setting? Informal learning in zoos (or anywhere) can take a multitude of forms. Unlike other settings, zoos offer learners a chance to receive information through all of their senses. Smell teaches the meaning of scent marking and territorial behavior in the big cats; hearing teaches the meaning of a mating call and its implications for species survival; vision explains adaptive coloration and describes the beauty of an emerald boa; touch conveys the roughness of an elephant's skin.

The definition of *informal* contains special meaning for zoo educators. It gives them freedom to be creative and to address the public in familiar, direct, and user-friendly ways. But any serious discussion of the zoo public and its needs calls for a much closer analysis of the visitor profile.

Who Is the Zoo Visitor?

Just about everyone is a zoo visitor, and this is the crux of the zoo educator's dilemma. School administrators who deal with a much more narrowly defined population often bemoan the impossibility of their task. After all, they say, providing quality education to individuals with a variety of socioeconomic, cultural, and educational backgrounds is exceedingly difficult.

But take a closer look at the typical Sunday crowd, and you'll understand the zoo educator's problem. Not only is there an enormous variety of people, but they span all age groups. A common misconception is that young children primarily visit zoos. Interestingly, a visitor survey conducted at the Bronx Zoo in the summer of 1986 revealed that 22 percent of visitors came to the zoo unaccompanied by children. Nationally, nearly 25 million adults flock to zoos annually without children in tow.

The lone adult phenomenon is far more prevalent than zoos expected. When the Bronx Zoo opened its renovated Children's Zoo in 1981, the media gave an erroneous impression that adults who were not accompanied by children would not be admitted. Stories were heard of many childless adults attempting to "borrow" children from their neighbors, relatives, and friends to be admitted to the wonderland where anyone could see a prairie dog or be a prairie dog, see a turtle or be a turtle. Never before or since have there been such generous babysitting offers.

A common practice of the New York City Board of Education demonstrates another example of the magical attraction of zoos. Before budget cuts, several fulltime truant officers were permanently assigned to the Bronx Zoo. Their job was to retrieve adolescents who planned to spend their day out of the classroom "playing hooky" at the zoo. What a tribute to the power of zoo exhibits as motivators for informal learning!

Let us take a closer look at the mass of humanity that pours in through the gates of every zoo. It seems reasonable to assume that zoogoers represent a segment of a normal population. One could predict that people at one extreme will be very active participants, seeking out courses, lectures, special tours, and avidly reading most graphic displays. An equal number of people at the other extreme will be passive, perhaps even disinterested. These people will learn solely from viewing an exhibit. The largest remaining segment of the audience has middle-of-the-road tendencies and is the one on which zoos must focus their attention. This group presents the greatest challenge as well as the greatest opportunity for zoo educators. In the United States alone, this group accounts for roughly 74 million visitors each year. The group probably ranges from people who are somewhat passive to those who are mildly active. Members of this group probably read some graphic displays, but only the most engaging ones; ride the camels, the monorail, or the safari trains, if only to rest their weary feet; and might partake of colorful special events, particularly those they find entertaining rather than overtly educational.

How can zoos best deal with this large, amorphous group of informal learners? Perhaps by breaking them down into smaller, more narrowly defined and consequently, more manageable subgroups. The business world frequently applies the concept of market segmentation to increase the selling power of various products. In some ways, zoos can benefit by borrowing the business model. The audience of zoo visitors contains many clearly identifiable subgroups even at a cursory glance. Zoos serve adults, children, families, couples, teens, schools, college students, camps, yuppies, all races and nationalities, seniors, and the handicapped. A closer inspection reveals even more distinct segments. We are likely to find white-collar and blue-collar workers, Gray Panthers, photographers, art lovers, teachers, tourists, scouts, single parents, and many other definable subgroups.

Perhaps by creating programming more closely linked to the needs and characteristics of these subgroups, zoos can achieve greater success in selling their product. But before we analyze these needs, we should carefully define the nature of the product. This is where the usefulness of the business model diminishes. Some might define the product as the exhibit or the live animal itself, but in a truly modern zoo this definition is incomplete.

As zoos assume an ever more critical role in wildlife preservation, they need to sell the ideas of conservation, biological diversity, population control, sustainable yield harvesting, species survival plans, and other unfamiliar concepts. Unlike businesses, zoos must market *ideas* rather than products. Conveying the value of a tangible product is a relatively simple task, but how can zoos convey the value of abstract ideas more effectively? This is the very heart of the zoo educator's dilemma.

Solutions for the Twenty-First Century Zoo

With some deeper thought, a few guidelines for "selling ideas" emerge. Ideas should be articulated clearly and directly. They should represent a philosophical consensus of the zoo profession and should relate to the visitors' lives. This is not always a simple task. Zoo educators are still embroiled in several debates.

Should tame live animals be used in zoos as teaching resources? Do Children's Zoos serve any useful functions? Do participatory exhibits really work, or does the medium become the message? Do animal shows demean the animal? Should more effort be expended on teaching youngsters or adults? How can U.S. citizens be made to care more about habitats in remote parts of the world? Is evolution an appropriate subject for zoo exhibits? Can the emotions be influenced through intellectual approaches?

Although the answers to such questions are still debated, educators accept some key concepts. For example, they agree that it is desirable to present ideas in some sort of a hierarchical sequence, proceeding from the more familiar to the complex, and building the most abstract constructs on a solid foundation of basics. They also know that they can make an idea more appealing by personalizing it and presenting it in terms of the visitor's immediate experience. Another important guidepost to selling ideas is to present them in layers of complexity, taking into account that zoo learners represent enormous intellectual and motivational variability.

There is no question that to be successful, zoo education programs in the next decade will have to be more closely linked to the diverse needs of zoo constituencies. Each audience segment needs its own special approach. Zoos will not be able to restrict programming to the groups most appealing to funding sources (i.e., the gifted, young children, the underprivileged). Rather, zoos will have to educate donors about the value of conveying ideas to the masses—a rainbow of people representing all ethnic groups, ages, and professions.

The necessary commitment to new programs will require considerable resources. Indeed, it is unlikely that the target audience will be either willing or able to contribute significantly to the cost of developing and providing such programming on a sustained basis. This is a particularly serious problem for zoos, which primarily are nonprofit institutions struggling to meet the cost of education while maintaining animal breeding and exhibition programs. Faced with such a challenge, zoos will need to marshall all the technologies currently at their disposal and establish cooperative "think tanks" to find the most effective use of their living collections for the benefit of distant wild habitats.

Of all the possible resources, live animals are the zoos' most valuable treasures not only because of each animal's intrinsic and biological value, but also because each living creature is endowed with untold educational potential. In the final analysis, the captive collections in the world's zoos will speak most convincingly on behalf of their wild and "free" brethren. Zoo educators and zoo exhibit designers must create informal learning opportunities that take full advantage of what each animal has to offer.

The Art and Science of Questioning

An animal's geographic distribution, ecology, evolutionary history, morphology, physiology, anatomy, and behavior are just some topics that lend themselves to interpretation. A clearer focus on any of these aspects also yields a wealth of additional areas for productive investigation. How does the animal cope with the climate of its habitat? How does it make or find shelter? What nutrients are essential for its survival? How does it obtain them? How does it navigate through its environment? Who are its competitors? How does it reproduce? How does it identify other members of its species? Does it care for its young? Is it social or solitary? Has its population remained stable or is it declining? What role do humans play in its future survival? These are just some questions curious and properly directed zoo visitors can explore.

Curiosity is the motor that drives the scientific pursuits of a zoo "explorer." But zoo designers and educators can engage the visitor in the scientific process by cleverly spotlighting subjects that merit study. Close observation is a first step. However, the great majority of urban zoo visitors have had little experience with this skill. Learning good observation skills takes practice, but a few basic questions could be placed at strategic locations to encourage careful observation. For example, visitors might be encouraged to determine how many animals are on display. With active or camouflaged species this task alone could turn into an enjoyable game. Furthermore, visitors might be asked to figure out which animals are male or female. This in turn may encourage closer observation of individual behaviors. Perhaps along the same line, viewers might want to consider the age of the animals. Having gone this far, some visitors might wish to pose hypotheses and conduct more observation, or data gathering, to determine if their guesses (hypotheses) were correct.

Clearly, not all visitors will pursue the scientific process through all of its steps. Time constraints, crowded exhibits, the animal's daily cycle, and other variables will interfere. However, the essential fact is that animal exhibits can open entirely new realms to the nonscientist. Properly exhibited, the living, behaving animal brings out the scientist in everyone. After all, aren't the acts of observing and asking questions the major activities of science? Don't scientists pose more questions than they answer?

Zoo educators as well as visitors need to develop the art and science of posing good questions. Because education is a process of imparting and acquiring knowledge, zoo educators and visitors can learn from one another. The diversity of animal lifestyles and their survival strategies are so endlessly fascinating that doing so should be a relatively easy task.

Just a quick glance at the discovery cards from the Bronx Zoo's recently published curriculum entitled WIZE (Wildlife Inquiry through Zoo Education) suggests a plethora of discoveries that you can make by studying animals. How might you be different if you had an exoskeleton? Why do the eyes of a chameleon move independently of each other? How do badgers keep their burrows dry during a flood? How are the teeth of a gorilla similar to your own? How do predators learn the meaning of warning coloration? How can the albatross stay aloft without beating its wings? Why do hummingbirds have a higher metabolism than other birds? How do deep-diving mammals avoid the bends? Why do some toads carry their young embedded in the skin on their back?

One important thing zoos can do to improve the understanding of each animal's role in the ecosystem is to present information that shows how the animal interacts with its physical environment and with other species of animals. Visitors can also explore the effect of human activity on animal survival. The interconnectedness of living systems must be an integral and recurrent theme of interpretive programs. Creative interpretation can also make room for connections between animal study and other areas in science.

Like other science disciplines, animal science can be integrated within a larger context of science, making it more dynamic and relevant. The Bronx Zoo is currently involved in a major effort to show a nationwide audience of teachers how to accomplish this kind of curricular integration.

For example, how many physics teachers would think of a trip to the zoo to explore concepts in physics? Yet a closer look reveals a wealth of opportunities. The study of optics in the vertebrate eye, the spectrographic analysis of feather pigments, transmission of vocalizations in the air and in the water, thermoregulation, calorimetry of various diets, the physics of bird flight, insulating properties of body coverings, buoyancy of aquatic animals, or mechanics of gibbon brachiation are just a few physics concepts that could come to life at the zoo.

Other science disciplines can also benefit from this kind of integrative approach. Chemistry teachers could conduct a fascinating chemistry lesson exploring the effects of acid rain on aquatic animals. Geography and geology teachers can make lessons more interesting by incorporating the study of patterns of animal distribution, both in modern and prehistoric times. The behavioral sciences such as psychology or sociology can also be enriched by including animal studies. Knowledge of patterns of social behavior, dominance hierarchies, aggression, and territoriality in animal species can enhance our understanding of the roots of human behaviors. The point needs no further belaboring, because for the creative instructor the zoo offers a gold mine of opportunities to motivate almost any learner. The living, breathing animal is better than any text, illustration, or laboratory experience for learning even the most sophisticated scientific concepts.

Considering that today's zoos have an entirely new role in species preservation, perhaps in the not-too-distant future zoo scientists can call on the lay, yet scientifically literate, zoo visitor to share in a partnership for species protection. Properly involved, the great masses of urban zoogoers can influence the course of animal preservation. By providing a base of financial, political, and moral support, this group can actively participate in the zoo's role as Noah's Ark. But this will not happen until zoo educators realize that there is much to be done to bring a modicum of scientific literacy to the average zoo visitor.

Marvelous and enjoyable though they are, zoos have yet to fully realize their potential as centers for informal science education. Some zoos today are already at the brink of the twenty-first century in the animal management realm. Modern technology and the grim statistics of species extinction have propelled them into a new age. Frozen embryos, ova and sperm banks, surrogate mothers, computerized matching of mates, and scientifically determined sex ratios and inbreeding coefficients are the stuff of science fiction rapidly becoming the daily reality. Zoo education must make a major leap to catch up with tomorrow.

Reference

Conway, William G. (1984–85). The practical difficulties and financial implications of endangered species breeding programs. *1984–85 International Zoo Yearbook.* (Vol. 24/25). London: Zoological Society of London.

Helpful Hints at the Zoo

- Don't expect to see lions and tigers and bears. Modern zoos exhibit fewer species in larger social groups than in the past. Therefore, not every species will be represented in your zoo's collection. Declining numbers of species on exhibit can be a positive sign that your zoo is modernizing its animal management practices. No longer is a zoo's quality measured solely by the number of species in the collection. To avoid being disappointed, try to find out ahead of time which animals are on exhibit.

- Do plan to visit a specific area of the zoo, particularly if you are visiting a large urban zoo. If you try to cover too much, you will miss an opportunity to make exciting observations. Take your time, don't rush, and keep in mind that animals don't behave on cue. Your patience will often be rewarded by an unexpected discovery: a sudden courtship display, a song, a flash of color, or even a birth.

- Don't look for specific animals. Modern zoos frequently exchange animals to increase the genetic diversity. Your favorite animal from a previous visit may have been shipped to a zoo in another state or country. Try to think of the benefits to the species as opposed to thinking of individual animals.

- Do try to learn something about the activity cycle of your favorite animal groups. If, for instance, you come to see a nocturnal animal during the day, don't expect it to be active.

- Do find out what endangered species your zoo has in the collection, and try to learn what breeding success it has had with this species. Breeding endangered animals is an important goal of most zoos. Appreciate their efforts when you see the young of such a species on exhibit.

- Don't be too disappointed when a particular house or exhibit is closed. The zoo could be renovating old-fashioned exhibits, or a newborn may require privacy with its mother for a few days. In either case give your zoo credit for making progress.

- Do plan to visit your zoo during the less crowded seasons. Most zoos are open year-round. Your chances for interesting observations are often better in the off season. You will probably be rewarded beyond your expectations. Think of a trip to a tropical exhibit in the winter as a great mini-vacation.

- Do find out what educational programs your zoo has for your particular interest group. More and more zoos have education curators who plan a variety of course offerings for children and adults.

- Do check out volunteer opportunities. Many zoos train a cadre of dedicated adults who work as docents, providing guided tours and participating in many other public services. If you are a teacher, your skills will be particularly welcome.

- Do become a member of your zoo. It needs the grass roots support of concerned citizens. Members often receive publications that keep them abreast of the latest developments as well as admission to special events, previews, and behind-the-scenes facilities.

10 Making the Most of the Zoo

Pegi S. Harvey and Debra Erickson
San Diego Zoo
San Diego, California

What can replace seeing a giraffe's 46-cm-long blue tongue reach out for an acacia branch, or watching elephants roll on their backs in the mud? The authenticity of these experiences makes them memorable. Presenting them through models or television simply doesn't command the same lasting effect.

Zoos offer an escape from the confines of the classroom. They are places where learning adventures can integrate the "real thing" into your curriculum—living, breathing, three-dimensional sights. These visits motivate, stimulate, inspire, and exhilarate. Students begin looking beyond their neighborhood to form a more global picture of the world. Not only do they become worldwide travelers, but they also experience nature first-hand by touching tamanduas, smelling sage, hearing howler monkeys, seeing snakes, and tasting monkey chow. A visit to the zoo increases observation skills and relates education to experience.

Most importantly, the feelings resulting from these encounters lead to a permanent commitment toward conservation.

A famous conservation quote reads, "In the end we will conserve only what we love, we will love only what we understand, we will understand only what we are taught." Before students climb the values ladder and make a conscious effort to conserve, a progression must take place: They are taught . . . they understand . . . they love . . .they conserve.

The foundation for creating a conservation-minded values system in students begins with a teacher. By discussing endangered species and their habitats, students begin to understand and to appreciate the importance of each organism in an ecosystem. Concepts such as extinction and habitat destruction are often remote, complex, and difficult to understand. Yet by coming close to animals and actually seeing and touching them, students form a personal bond. Through time, this bond turns into commitment to care for animals and a realization that all species on Earth are interdependent.

By thinking globally and acting locally, students' commitment evolves into stewardship. They begin to realize that by recycling paper and conserving water they may be helping an endangered species. Conservation then becomes a more concrete concept they can act upon and commit to.

You can arouse students' curiosity about animals and plants until they're hungry for more knowledge. This not only affects their conservation ethics, but also acts as a catalyst for changing other behavior outside the classroom. Students begin making informed, conscious decisions—from choosing to read and watch nature-related material, to spending more time outdoors exploring the world around them. These decisions can involve the whole family, increasing the scope of influence.

So break down the barriers of the textbook, lab guide, and classroom. Use the zoo as a dynamic learning resource.

But I Don't Know Any of Those Plants or Animals!

You don't need to be an animal expert to teach creatively and effectively at a zoo. Zoo educators help bridge the gap between you and the plants and animals, so you can do the same with your students. But how do you get started?

First, call your zoo and ask them to send a brochure that explains their programs. Many zoos offer programs that include:

- guided tours
- self-guided tours
- self-guided tours paired with a puppet show
- slide shows and animal shows
- programs for gifted students
- special education programs
- career presentations
- outreach programs

After making reservations, many zoos send an information packet containing a "recipe" for your visit, along with pre- and post-visit activities. What happens if a curriculum aid isn't available, you need more information, or you're not sure what to do with the information you've received? Call the zoo! Many zoos offer a wide variety of teacher orientations and workshops—some are offered for credit through local universities. These offerings are designed to demystify the tremendous amount of information available and to reduce the planning burden. At some zoos, you can make an appointment to consult with a zoo education specialist to tailor a program to meet your needs.

Many zoos offer teacher orientations or specialized weekend training programs. During the summer, Monterey Bay Aquarium in California offers a 10-day workshop called Marine Science Teacher Institute for K–12 teachers. Teachers learn how to develop a marine science curriculum and take home not only specimens, slides, experiments, and worksheets, but also confidence in their ability to teach the subject. But their experience doesn't end there. Follow-up sessions help Marine Institute graduates discuss their successes and failures with fellow teachers.

Chicago's Lincoln Park Zoo and the San Diego Zoo offer lecture series on topics such as "Great Apes" and "Vanishing Species." After attending the series, a special class helps translate the information into classroom activities. High school teachers can send their top students to these presentations, and these students can later share their feelings on the lecture with their classmates.

You can also become involved with the docent program at your zoo. Docent programs teach individuals about ecosystems and their inhabitants along with techniques for presenting the information. Docents are assigned to certain areas of the zoo or asked to give tours to visitors and education groups. Docent training programs run 4 to 16 weeks and are usually offered only once or twice a year. The Lincoln Park Zoo has a short training program each spring so teachers can get involved as summer volunteers.

Teachers who lead summer school classes at the San Diego Zoo also come away with a new understanding on how to use the zoo. An education programmer trains teachers about the information and presentation styles. Teaching courses and explaining animals on a formal basis make it easier to transfer your respect and admiration for the animals to your students. You'll be amazed how many of your experiences you'll share with your class. The excitement and wonderment are contagious.

Other resources are available to prepare you for your zoo visit. Natural history museums often have compatible curriculum materials. Many have materials such as skulls and skins you can check out to prepare your class before your visit. Libraries are another resource. Ask the librarian at your school, district media center, city, or university library to point you in the right direction for books, films, records, and videotapes.

Project Wild and Outdoor Biological Instructional Strategies (O.B.I.S.) both offer workshops that show activities you can tie into a zoo visit. Use information from all these sources and call other teachers who have visited the zoo. Ask them for activities that worked well for them and try them out. Take time to investigate all the possibilities.

Charting a Successful Course

The most successful visits to zoos are those that are focused. It's impossible to see every animal and to talk about every ecosystem on your trip. Don't transfer the "cover the whole textbook" syndrome to your zoo visit. It will frustrate you and your students. Many zoo offerings have themes such as adaptations, behavior, African animals, animal care, animal locomotion, endangered species, and ecology. Choose one of these themes or create your own.

Since most teachers aren't accompanied by a guide when they visit the zoo, preparing yourself and your class is vital. But don't let that discourage or intimidate you. Guides aren't the only people who can unlock the zoo's secrets. Remember, you have an advantage in knowing your curriculum and your students' needs and talents. Stimulating and motivating teaching at a zoo doesn't depend on quoting facts and relating concepts. Rather, it relies on your ability to ask questions that actively involve your class.

Why is this involvement so important? Students retain about half of the concepts they encounter and even fewer facts and terms. Lectures that give long lists of facts are not effective and compete with the animals for your students' attention. But activities and attitudes have a great impression on your learners, so concentrate your efforts in those areas.

After deciding your theme, use available resources to prepare your students for their visit. Many teachers get parents, administrators, and other teachers excited about the trip, but they forget to transfer those feelings to their students. Get your students to look forward to their trip. Discuss the theme, do theme-related activities, and have your students compose a list of questions they have about the topic. Have them try to discover the answers to their questions during their visit.

Make your trip the hot topic of conversation at all the school's lunch tables.

Many zoos have a free planning pass you can use to visit the zoo and develop your trip's strategy. When you get home, make a list of all the things that excited and fascinated you. Include these things in your trip, because if they captivate you, they will certainly interest your students. Break down your plan into "chunks" that the students can easily digest. Yet don't treat the animals as postage stamps; make sure students can envision them in their ecosystem.

Remember you are facilitating a meaningful experience between the plants and animals and the students. Promote the concept of wonder, amazement, and diversity of life. Give your students time for uninterrupted observation so they can develop their own conclusions to questions such as why elephants flip sand on their backs or whether a zebra's stripes are black on white or white on black.

Carefully read all the instructions the zoo sends you so you can have a hassle-free admittance. Make sure you communicate to your chaperons what their responsibilities are. Give them a packet containing copies of zoo information along with the schedule for the trip and the list of rules for the visit. Follow your plan, but be flexible. If an animal is being fed or giving birth, stay and watch the event and cut out something else. If a guide is escorting your group, team up with them to maintain class control.

Don't let your visit be an isolated experience. Discuss what you learned after the trip. Give the students time to share their impressions and feelings. Discuss what amazed them along with their questions that didn't get answered. Involve them in post-visit activities such as writing poems or essays, or creating a mural or scientific drawing. Show them avenues for further investigation, such as reading material, nature shows, or nature trips offered by the Audubon Society or the Sierra Club.

Don't underestimate what you can do when you team up with the zoo. You will not only have a lasting impact on today's students, but also on the future of the plants and animals that share our earth. We need you as a member of our team.

Helpful Hints: Taking Your Zoo Experience One Step Beyond

- The San Diego Chargers sponsored a program called "Kicks for Critters." Individuals pledged donations for each field goal made for the season. Lee Childs, a math and science teacher, took this one step beyond. She created "Cans for Critters." Students collect aluminum cans throughout the year, usually about 83,000 cans, resulting in a $600 to $800 yearly contribution that goes toward research on endangered species.

- Following their trip to the Monterey Bay Aquarium, Vickie Madigan's communicatively handicapped class of fourth- through sixth-graders was so impressed with their visit that they turned their classroom and hallway at Robert Down School into an aquarium. Shadow box aquariums lined the hallway and each child was ecstatic about the exhibit he or she created. Some exhibits were recreations of the kelp forest complete with divers; others represented the otter exhibit or Monterey Bay Habitats exhibit, complete with schooling fishes and sharks. Students drew their favorite marine plants and animals to add to a huge mural labeled "Our Aquarium." Vickie said the experience was a great catalyst for improving communication for these special students.

- At the Atlanta Zoo, teachers and students worked with the help of the *Atlanta Journal Constitution* newspaper to raise money to buy two new elephants. By having bake sales and other activities, students raised more than $50,000 in two months. All the students involved in the project were invited to the elephants' unveiling. The four- and five-year-old African elephants are now in their new home.

11 The Science Museum: Object Lessons in Informal Education

Michael Templeton
National Science Foundation
Washington, D.C.

Why is a museum like a school? Like the Mad Hatter's "Why is a raven like a writing desk?," the question is more for argument than for answer.

Education by Choice

Why are museums informal educators? The audience is self-selected and the visit is freely chosen. Visiting a museum is an exercise in free choice, filled with incentive and opportunity for personal variation. The need to attract and hold an audience has led museums to combine entertainment and recreation with education and information.

The economics of free-choice learning makes museums sensitive to individual interests and needs. Individuals assemble their personal curriculum as they proceed through exhibits on diverse subjects that they can pass or view in varied order. (In a few museums, textbook-like linear sequence exhibits allow no choices; visitors respond by passing crowded or unappealing sections, or by swimming upstream against the current.)

Each of us learns differently. Museums provide varied, multisensory approaches to learning: objects, models, motion or animation, voice and music, touch or physical manipulation, visual images or video sequences. Accompanying text may be multilevel for different reading or interest levels, or multilingual for different cultural backgrounds. Science centers employ "explainers" instead of guards to help visitors understand the function and meaning of individual exhibits. Museum exhibits, however comprehensive they may seem, are highly selective experiences. Like most informal science education materials, whether a television program or a kit of science experiments, a museum exhibit is heavily edited by its creators and "self-edited" by each individual who experiences it.

Museums, unlike schools, reach large numbers of people—a given exhibit "curriculum unit" may be seen by 100,000 to 1,000,000 people a year. Museums employ unusual or expensive equipment, rare specimens, or painstaking research. Few schools can afford the cost and complexity of a large marine mammal display in an aquarium, an elaborate and realistic reconstruction of a tropical rainforest, the translation of a complex laboratory experiment into a science center exhibit, the elaborate, dramatic presentation of a large planetarium, or the ever-favorite reconstructions of dinosaurs in a natural history museum.

The unique character of museums is their reliance on objects and phenomena as the fundamental basis for education and communication. Although museums occasionally exaggerate the inherent learning power of specimens, models, artifacts, or experiments, these things are the living heart of museums.

Such objects both allow and require different ways of learning. The three-dimensional and visually rich environment of the museum gives new opportunities to those who do not think readily in textbook terms or who need objects to understand abstractions. Important ways of knowing about the world cannot always be represented in the flat world of blackboard, textbook, or television.

Exhibits allow individual attention by individual learners; they allow decentralized, self-paced learning. Object-rich museums can meet different students' needs at the same time. The learner receives feedback and reinforcement from touch, smell, sound, the reward of a response, or the more subtle interaction of a complex experiment whose parameters can be varied at will. At one time, museums' professional concern for preservation and care of unique objects obscured these important aspects of object-centered learning. Led by science centers, museums are once again recognizing the importance of interaction.

The Origins of Museums as Informal Educators

In refining the ancient practice of object-based learning, museums developed an approach to education and learning quite different from that of schools.

Science museums began in Europe between the fourteenth and seventeenth centuries as private cabinets of curiosities—social and intellectual constructions of and for the nobility. The strange, unique, and remarkable were the objects of choice. As late as the eighteenth century, museum attendance was limited to those with carriages and calling cards who made appointments for tours led by the founder or a knowledgeable servant. Prior to the work of such organizers of natural knowledge as Hutton, Linnaeus, Lamarck, Cuvier, and ultimately, Darwin, collections were fragmented and reflected the natural history of the time. Guides did what they could to structure the visit, pointing out objects of note, adding local color, and occasionally emphasizing a point of science or of natural history. As in society, children were absent or invisible.

The slow transition to public museums began with the opening of the Ashmolean Museum at Oxford in 1683 and the British Museum in

1759—a transition that would take another century to complete. As eighteenth-century science became better organized and societies and journals flourished, science also became a subject of popular interest among a rising middle class. Privately operated museums began to compete with the British Museum. Scientific knowledge developed disciplines and divisions; geology and biology became separated from the "physical" sciences, and museum collections became classified by subject.

Near the end of the eighteenth century, the American and French Revolutions marked the beginning of the democratization of life. Museums felt the impact; the Revolution led to the founding of the Museum National d'Histoire Naturelle (Paris) in 1793 and the Conservatoire Nationale des Artes et Métiers (Paris) in 1794.

In England, a movement for popular education in crafts, technology, and the underlying sciences arose in response to the social impact of mechanization and industrialization. The Royal Institution (London), founded in 1800, initially emphasized adult, informal science education. Its director, Sir Humphrey Davy, made popular science lectures significant London social functions. His successor, the great physicist Michael Faraday, began a program of popular Friday evening lectures in 1826, and instituted the first of 19 annual "Christmas Courses of Lectures Adapted to a Juvenile Auditory." These lectures established the lecture-demonstration as a polished means of popular communication and legitimized informal science education for the young.

A parallel demand for broad public access to museums for culture, education, and recreation resulted in regular hours for the British Museum in 1807. "Any person of decent appearance" was admitted without an appointment. A heated public debate in 1837 led to the opening of the museum to the public on holidays; on the first Easter Monday, more than 23,000 people came.

Popular exhibitions added a populist flair. Giant paintings, landscapes, or dramatic scenes of battle as large as 20 feet tall were popular at the end of the eighteenth century. Great panoramas and dioramas followed in the early years of the nineteenth century. Whole buildings were devoted to realistic portrayal of popular scenes or giant paintings on moving rolls of canvas, and were equipped with machinery, music, and all the tricks of stagecraft. These exhibitions drew large crowds at great expense. In subdued form, these techniques led to natural history museum dioramas that represented natural life and to the atmospheric effects of planetarium light and sound.

Practical models of productive technology—machines, mechanisms, boats, and fishing gear—introduced the "working classes" to museums. The Adelaide Gallery (London, 1834) and the Polytechnic Institution (London, 1838) were exhibition centers remarkably like modern science-technology centers. The Polytechnic Institution offered exhibits, demonstrations, lectures, and teacher training activities, and charged a shilling entrance fee. As audiences expanded and broadened, exhibit guides became guards, and explanatory and expository text were added to museum displays.

Finally, the Crystal Palace Exhibition of 1851 created the first great world's fair of technology and commerce, which exalted industry, its machines and products, and technology and progress. In the United States, early museums borrowed European styles with the Peale Museum (Philadelphia, 1786) and P. T. Barnum's American Museum (New York, 1841). These early American museums were informal and popular and the educational nature of their contents widely advertised. Science museums developed with the first of the Smithsonian's museums (Washington, D.C., 1856), the American Museum of Natural History (New York, 1869), and the influential Centennial Exhibition of 1876 in Philadelphia.

In Germany, the idea that museums could represent working examples of machines and technologies—that objects could explain themselves through their operation—was developed by

Oskar von Miller in the Deutsches Museum (Munich, 1906; 1925). Julius Rosenwald adapted this idea at the Museum of Science and Industry (Chicago), which after its opening in 1933 rapidly became America's most popular museum of industry and technology. The Franklin Institute Science Museum (Philadelphia, 1933) and the Buhl Institute of Popular Science (Pittsburgh, 1939) also reflected the rising interest in this country for popular, informal education in science and technology and the creation of what would come to be called science-technology centers.

The most dramatic expression of this new emphasis on science, rather than on collected objects from history or nature, came with the transformation of Boston's established museum of natural history into the Museum of Science immediately following World War II. Educating the public about science and technology became the main goal of the new museum, and prior commitments to collecting, preserving, and studying natural history were replaced by a new emphasis on informal science education.

More than 100 new American science centers have been founded since the end of World War II with informal science education as their primary mission. The participation of America's scientific community in developing the U.S. Science Pavilion (now the Pacific Science Center) at the 1962 Seattle World's Fair reflected this new attention to informal science education. Two remarkable years saw the opening of the Ontario Science Center in Toronto in 1969; the Lawrence Hall of Science in Berkeley, California, in 1968; and the Exploratorium in San Francisco in 1969.

In response to the success and popularity of these and other science centers, an international movement has developed with the goal of creating new science museums dedicated primarily to informal education. A tremendous resurgence of interest in their educational mission has led natural history museums, zoos, and aquariums to adopt some of the appropriate techniques of the science-technology center.

Museums and Schools: Contrasts and Consequences

How then do museums and schools differ? If the differences between formal and informal learning result in unique roles for each, then what do museums do that no others do as well?

Schools are predominantly systems of education, with each school, grade level, or class a well-ordered element of an educational whole. Maintaining the system—economics, behavior, and pedagogy—dominates. Knowledge and instruction generally appear as continuous and sequential, purposefully selective, explicated by textbooks, and organized and managed by lists of necessary instruction (scope and sequence documents) and necessary performance (standardized performance on standardized tests).

As students reach high school and beyond, knowledge derived from scientific research is even more tightly structured, precisely formulated, universal, mathematical, and theory-bound. The process of learning is represented to the student as a responsibility, a requirement, and a duty; all under the unfair heading "formal education."

Learning outside this orderly context is called "informal education," which means learning that centers on individuals who freely choose whether, when, and what to learn. Sources for informal science education are diverse and disaggregated. They share an intense concern for the attention of the individual learner; exploitation of curiosity, novelty, and occasionally, the bizarre; compromises between the complex content of science and a necessary simplicity and ease of understanding; and utter disregard for the distinction between learning as enjoyable personal recreation and education as a serious, responsible business. Modern science-technology centers and science museums are a primary institutional base of informal science education. They combine the entertaining aspects of visiting museums and exhibitions with object-based and participatory learning; this is why they are popular and effective adjuncts to schools and libraries.

In contrast to formal education in schools, a science museum's informal education, expressed through public exhibitions and programs, is highly variable and selective, edited according to rules and values idiosyncratic to each museum. Essential knowledge is represented by things rather than texts—collected specimens in natural history museums, constructed exhibits that display or model scientific phenomena in science centers. Small museums are highly selective, while large ones contain so much material that it cannot be seen in a single visit. Consequently, visitors perceive museums, whether rightly or wrongly, as complex, branching, and nonsequential experiences, with few interrelated components.

Museums do not meet the many requirements of classrooms or laboratories. Museums do not guarantee that participants will master subjects. Order and decorum—control over individual behavior—is minimal by comparison. There are virtually no measures of knowledge gained or proficiency attained. And yet, museum visits are highly prized, sought-after experiences that deeply stir students and teachers alike. People recall museum visits with fondness and significance years later. At their educational best, museums are surprising, exciting, and engaging, while lacking the predictability and thoroughness of schools.

Museums represent discovery; they are libraries turned inside out. The museum visitor has no catalog of contents available and no recourse other than a brisk walk through the halls and galleries.

A museum's contents announce themselves and are a real, tangible, representation of ideas. In general, one exhibit refers to another only by juxtaposition; the order and manner of use (and thus instruction) is largely up to the individual.

Learning directly from things is deeper, more primitive, and more fundamental than learning from formal lectures, texts, and lesson plans. But the pedagogy of informal, object-based museum learning is far less well understood. When we enter a museum, we enter the age of Homer, when objects were themselves powerful, memory a principal tool of scholarship, and songs and ballads a means of instruction. Our present ability to create exhibits of things from which we learn is similarly primitive and similarly Homeric—passed down from one practitioner to another by example, rather than from text to text.

Informal Learning in Museums: Conclusions

If the museum experience is primarily the experience of things, how does a visit to a museum differ from everyday experience, or from the instructional use of objects in a classroom? Why not make all of nature, art, and environment our museum? If museums are simply reductions of our experience, why are they so stimulating and exciting when we arrive?

First, museums reflect the limitation and circumscription of their contents. While museums are never as disciplined and organized about their collections as they would claim, even the nest of a magpie or a packrat is remarkable for the concentration and juxtaposition of its objects. As seen by visitors, museums share with dumps and midens the magic of uninformed accumulation. Things take on meaning and value when held out for contemplation, regardless of their organization.

Second, at their best, science museums contain carefully edited collections of objects, whether natural history, physical phenomena, or science's art. They are classified by some scheme and sifted by some notion of beauty, craft, or place in a universe of knowledge. They are selected and select examples of our intellectual culture.

Third, they are salient. Exhibits are designed to relate to key concepts, ideas, or components of a subject area. They remind us of what we want to know or what we want to remember about a subject. Exhibit components honor by their selection the significance of particular content. They are exemplars of knowledge and of a pedagogy that provides instruction in that knowledge. Practitioners of science and technology assemble exhibits to represent a curriculum with the objects and experiences of the museum. This curriculum cannot be mistaken for that of the school, however, for it is shaped by the unique necessities of exhibition display and the subtle character of object-based learning. It is particular, never general. It is rarely comprehensive, never encyclopedic. It is a curriculum none the less.

Fourth, museum objects and exhibits are careful constructions of sophisticated scientists, educators, craftpersons, and tinkerers. Like well-crafted lessons from a master teacher, there is much that does not meet the eye. Many failures and false starts, many notions that did not work have been tried and cast aside. At their best, they create experiences never seen in the classroom; the library of experiments in visual perception at the Exploratorium (San Francisco) are qualitatively and quantitatively superior to any other source. As constructions they have the point of view of their creators; they are neither neutral statements of fact nor self-evident explanations of "nature." At their best they have the selectivity and clarity of great essays.

The "modern" science museum, whether science-technology center, natural history museum, or nature center, reflects a tradition of museum practice that is interwoven with the development of both science and science education. To use museums effectively, one must recognize the unique character of museums that makes them educators parallel with but independent of formal education. In moving from school to museum and back again, the task is to make the most out of their different ways of education. A field trip to the museum is far more than escape from the classroom. And the return to school should be marked by a renewed interest in the processes and subjects of school science, rather than a return to the humdrum and ordinary. When this happens, the museum is not simply an entertaining and duplicative luxury, a good excuse for a break. Instead, it becomes a necessary place of informal education and learning, filled with essential opportunities not duplicated in field, home, or classroom, and it engenders a lasting influence on those who participate.

For Further Reading

Alexander, Edward P. (1983). *Museum masters.* Nashville, TN: Association for State and Local History.

Alexander, Edward P. (1979). *Museums in motion.* Nashville, TN: Association for State and Local History.

Altick, Richard D. (1978). *The shows of London.* Cambridge, MA: Belknap Press.

Danilov, Victor J. (1982). *Science and technology centers.* Cambridge, MA: MIT Press.

Lucas, A. M. (1983). Scientific literacy and informal learning. *Studies in Science Education, 10,* 1-36.

Tressel, George (1984). A museum is to touch. *1984 yearbook of science and the future.* Chicago, IL: Encyclopedia Britannica.

12

The Muse in the Classroom

Joel N. Bloom
Franklin Institute Science Museum
Philadelphia, Pennsylvania

Architect-philosopher Buckminster Fuller described children as "born true scientists," full of curiosity and the urge to investigate. This natural curiosity can be nurtured, along with the understanding that science is relevant to daily life. If children and teachers become partners in experimentation and exploration, neither is susceptible to the unfortunate impression that science is boring, irrelevant, and incomprehensible.

The Franklin Institute Science Museum in Philadelphia has been quite successful in developing partnerships to improve science teaching in the schools. Indeed, all over the country, school districts are turning to museums for collaborative efforts in science education and teacher training. In Philadelphia, a support and advocacy group called the Committee to Support Public Schools and the Philadelphia School District have launched a major science and mathematics initiative: PRISM. Strongly supported by the area's leading corporate executives, the Philadelphia Renaissance in Science and Mathematics seeks to strengthen science education by nurturing connections between the private sector and school district, museums, and other educational institutions. A particular concern is to increase the participation and achievement of minority students in higher level science and mathematics courses.

Museum-to-Go

To date, the Franklin Institute's most ambitious project is Museum-to-Go, a program developed over the last several years with the support of the Committee to Support Public Schools, the Atlantic Richfield Foundation, and the Gulliam H. Clamer Foundation. Museum-to-Go is just that: It brings the hands-on learning of the Franklin Institute into the classroom. It provides complete, easy-to-use hands-on science kits for elementary and middle school students, as well as workshops that train teachers how to use these materials. Teachers, administrators, and supervisors have been involved in every aspect of kit and workshop development.

The kits correlate with the standard elementary science curriculum and emphasize the physical and earth sciences. Each kit is a self-contained resource package for an entire class to conduct key experiments on a given topic. Since each kit is complete, the teacher does not waste valuable time seeking and organizing these materials. Classroom implementation is immediate. The kits are available on a loan basis, and as each kit is used in the classroom, it is returned for refurbishing so the next teacher will receive a classroom-ready package. These kits are available only if teachers attend the appropriate training workshops.

A Museum-to-Go training workshop is a stimulating, nonthreatening experience where teachers exchange ideas and brainstorm with colleagues from other schools. Each workshop covers a different topic, focusing on four interrelated dimensions: How To Teach Science, What To Teach, Why Teach Science, and How To Evaluate Learning. Typical units cover chemistry; acids, bases, and salts; energy; and meteorology. The 3,000 teachers who have participated in Museum-to-Go workshops welcome the opportunity to address their own believed inadequacies in science: their lack of familiarity with basic science, lack of training in experiential science education, and lack of confidence in their own abilities.

Thanks to the unstinting cooperation of the teachers and administrators of the Philadelphia School District, Museum-to-Go has surpassed all expectations. Our training workshops are always oversubscribed. Teacher evaluation has been overwhelmingly positive. All of the teachers who have participated say they would recommend Museum-to-Go science kits to other teachers.

Students are also enthusiastic supporters of the program. Nothing in my career has made me more proud than the letters from students who attend the Moffett School in North Philadelphia. Excited about science and asking for more, Maria, a sixth-grader, wrote, "I told my mother that science is my best subject. I never knew I could do science. Now I know I can." Many of the school's students, who represent 26 nationalities, do not speak English well. Only 10 percent are native English speakers. Even so, all students participate fully in science because the school's administration is committed to hands-on science education. As a result, the fifth- and sixth-grade teachers took our workshops and now use our hands-on activities. When one of our staff visited the school, two Asian children guided her through their classroom, pointing out their electricity experiments and proudly explaining in broken English the difference between parallel and series circuitry. One sixth-grade class constructed a doll house, fully wired.

The Museum-to-Go concept was inspired in part by earlier research work conducted at the Franklin Institute. Borun et al. (1983) showed that students learned more when their classroom experiences were enhanced by a museum visit. Perhaps even more significant, they *enjoyed* the learning experience, found it more interesting, and were motivated to keep learning. The study concluded that science is more accessible to children when classroom learning is supplemented by the hands-on experience of a science museum. This conclusion inspired us to create a program that made hands-on science a regular classroom experience, not just a once-a-year field trip.

The hands-on learning of Museum-to-Go is especially significant for an inner-city school system. Economically and academically disadvantaged students derive the greatest educational benefit from direct interaction with science activity materials, according to studies of test results from 13,000 students who participated in activity-based science programs (Bredderman, 1983). Experimental activities do not rely heavily on reading skills, and many disadvantaged students are poor readers. These children have been known to succeed in science classes at a rate comparable to those with better language skills.

In fact, publications by the National Science Teachers Association (Mechling & Oliver, 1983) conclude that science activities seem to improve communication skills as well. At the elementary level, activity-oriented science programs improve the academic skills of students who are currently "at risk," and may well increase minority representation in the science profession.

The Philadelphia Kit Collaborative

The Philadelphia Kit Collaborative builds on the success of the first four years of Museum-to-Go. By 1992, Museum-to-Go kits will be introduced to 2,800 Philadelphia elementary school classrooms. At each elementary grade level, four hands-on science activity kit units will fulfill four major curriculum goals.

The Philadelphia Kit Collaborative is the most comprehensive program of its kind in the United States. It is a national model for ways that science museums, concerned corporations, advocacy groups, and major urban school districts can unite to improve public science education.

Summer Institute

Each summer, 40 exemplary elementary teachers, science specialists, and principals receive indepth training in hands-on science education, using all the resources of Museum-to-Go in conjunction with the museum itself and our staff of educators. Funded by the U.S. Department of Education, the Summer Institutes help 10 Philadelphia elementary schools become science demonstration sites. Once again, the Philadelphia School District and PRISM have joined the Franklin Institute to improve science education, particularly at the elementary level.

Camp-In

Not all our programs are as formal as the Philadelphia Kit Collaborative or the Summer Institute. Ask 450 elementary and junior high school teachers from Pennsylvania, New Jersey, and Delaware who participated in the First Annual Teachers' Camp-In in 1986.

Camp-In is a well-established offering at many U.S. science museums. Founded at the Center for Science and Industry in Columbus, Ohio, in 1972, Camp-In offers a unique experience to youth groups—Girl Scouts, Boy Scouts, and others. Sleeping bags in hand, campers actually camp inside the science museum, enjoying exhibits and special programs during this overnight adventure. Camp-In has provided unusual educational experiences to thousands of young people in the past decade.

The Franklin Institute operates a successful Camp-In program for youth groups, but we also pioneered the Teachers' Camp-In. At the first Camp-In, teachers participated in earth science and astronomy workshops, listened to Ira Flatow, host of PBS' *Newton's Apple*, experimented with educational materials and computer software, and explored the museum itself. Rumor has it that one or two teachers actually slept a bit. Most stayed up all night and watched the sun rise from our rooftop observatory. "The camp-in changed my whole focus on teaching science," said a second-grade teacher. "You can't teach science from a textbook." We started receiving inquiries about the Second Annual Teachers' Camp-In before the first Teachers' Camp-In was over.

Energy for the Fifth Grade

In Philadelphia, energy and electricity form a large part of the fifth-grade science curriculum in both public and diocesan schools. In the 1985-86 school year, every fifth-grader in Philadelphia's public and diocesan schools—nearly 20,000 students—visited the Franklin Institute for a special energy-themed program, Energy for the Fifth Grade, designed to complement the fifth-grade science curriculum. We chose the fifth grade because middle school is a traditional time for field trips to the Museum, and because of the convergence between the science curriculum, the educational resources of the Franklin Institute, and established attendance patterns.

Recognizing the logistical realities of a museum visit for inner-city school children, Energy for the Fifth Grade provided free transportation (thanks to generous support from Philadelphia's William Penn foundation). Students attended a 45-minute presentation in our Science Auditorium that featured dramatic props, including a working airplane engine. This lively, participatory show explored heat, light, sound, and mechanical energy; demonstrated potential and kinetic energy; and showed how energy is transformed from one form to another. A special tour of the museum guided students to exhibits on energy-related topics, and an Energy Exploration exhibit invited them to find answers to energy-related questions.

Energy for the Fifth Grade was a multiphase program. We distributed to all participating classrooms a resource book that contained pre- and post-visit activities that explored the basic themes of the program: energy sources, types, and transformations. Students prepared for their museum visit, and afterwards, discussed what they had learned and experienced.

Corporate Contribution

Bell of Pennsylvania underwrote a three-year program with the Franklin Institute for two schools (Dunbar Elementary and Wanamaker Junior High) near its new corporate computer facility in economically depressed North Philadelphia. The program includes science museum visits and hands-on computer lessons, as well as visits from the Institute's popular Traveling Science Show. This program will not only improve science education in these schools, but also contribute to a good relationship between Bell and the residents of the new neighborhood.

Field Trips

All of these programs, from Museum-to-Go to the Bell program, are relatively structured. Another key component of the Franklin Institute–school relationship is the more familiar, less structured school visit to the science museum. Every year, more than 150,000 school children visit the museum on class field trips. They come to experience star shows in the Fels Planetarium, enjoy lively programs in the Science Auditorium and the Discovery Theatre, and participate in one of the many special classes presented by our museum education staff. Another 350,000 students experience our Traveling Science Shows every year at their own schools.

Publications

We have also produced three educational publications for classroom use, in collaboration with the Newspapers in Education project of Philadelphia Newspapers, Inc., publishers of the *Philadelphia Inquirer* and *Daily News*. Distributed as supplements to the newspapers, these 16-page publications have provided educational material for nearly 100,000 students to date on topics such as Halley's Comet, Women in Science, and Hands-On Activities for the Elementary School Classroom. A fourth supplement will be published in the autumn of 1988.

The Franklin Institute Science Museum has

provided informal learning experiences for the three-quarters of a million people who visit us each year, as well as more structured programs for schools. This tradition has provided entertaining hands-on science education to more than 40 million people since we opened our doors in 1934. The Franklin Institute also provides a laboratory to develop and test educational techniques and devices. At the same time, our formal partnerships with schools grow broader and deeper as we seek to find new ways to apply our expertise in experiential, hands-on education in the service of improved science education in the classroom.

References

Borun, Minda et al. (1983). *Planets and pulleys: Studies of class visits to a science museum*. Washington, DC: Association of Science-Technology Centers.

Bredderman, Ted. (1983). Effects of activity-based elementary science on student outcomes: A quantitative synthesis. *Review of Educational Research*, *53*(4), 499.

Mechling, Kenneth R., & Oliver, Donna L. (1983). *What research says about elementary school science*. Washington, DC: National Science Teachers Association.

For Further Reading

Apelman, Maja; Hawkins, David, & Morrison, Philip. (1985). *Critical barriers phenomenon in elementary science*. Grand Forks, ND: University of North Dakota Press.

Bennett, William J. (1986). *First Lessons: A report on elementary education in America*. Washington, DC: U.S. Government Printing Office.

Bennett, William J. (1986). *What works: Research about teaching and learning*. Washington, DC: U.S. Government Printing Office.

Cohen, Rosalie. (1986, November). A match or not a match? A study of intermediate science teaching materials. *Forum '86: The Science Curriculum*. Paper presented at the National Forum for School Science.

Hawkins, David. (1983). Nature closely observed. *Daedalus*, *112*(2).

Lapp, Douglas M. (1980). Helping teachers teach science: The need for teacher support systems. *The National Elementary Principal*, *59*(2), 61.

Linn, Marcia C. (1986). *Establishing a resource base for science education: Challenges, trends, and recommendations*. Berkeley: University of California Press.

National Commission on Excellence in Education. (1983, April). *A nation at risk: The imperative for educational reform*. Washington, DC: U.S. Government Printing Office.

National Science Resource Center. (1986, September). Position paper presented at the National Conference on the Teaching of Science in Elementary Schools, July 13-16, 1986, Washington, DC.

Helpful Hints

- Call the museum and ask if reservations can be made. Reservations may be required for special programs, guaranteed lunchroom space, or to receive a discounted group rate.
- Call the museum's Education Department several weeks before your planned visit. Ask for information about the exhibits and about special educational programs that may complement your science curriculum and enrich your visit. Many museums offer printed material that will also be useful; ask what is available.
- Visit the museum in advance. The Franklin Institute offers a free Educator's Pass to help teachers prepare for their visits; most other museums offer free or reduced admission to teachers.
- Don't try to do it all. It is impossible to see and

do everything in a single visit and have an educationally worthwhile experience. Concentrate on a few exhibit topics.

- Prepare your class with activities that relate to the exhibits you'll be visiting. The museum's Education Department can suggest pre-visit activities.

- Give your students specific assignments and review them in advance, or divide your class into teams and have them work together on an assigned topic. An assignment might be a question to answer or observations to make at the museum, or finding and preparing a classroom activity that relates to a topic you'll be exploring at the museum.

- Let your students help plan your visit, including rules for appropriate behavior. Using the museum's floor plan, plan your route to the exhibits. It will be a good exercise in mapping.

- The museum staff is there to help. Ask for directions, explanations, and other assistance.

- Consolidate your gains. Have a post-visit session to review what you've observed and experienced.

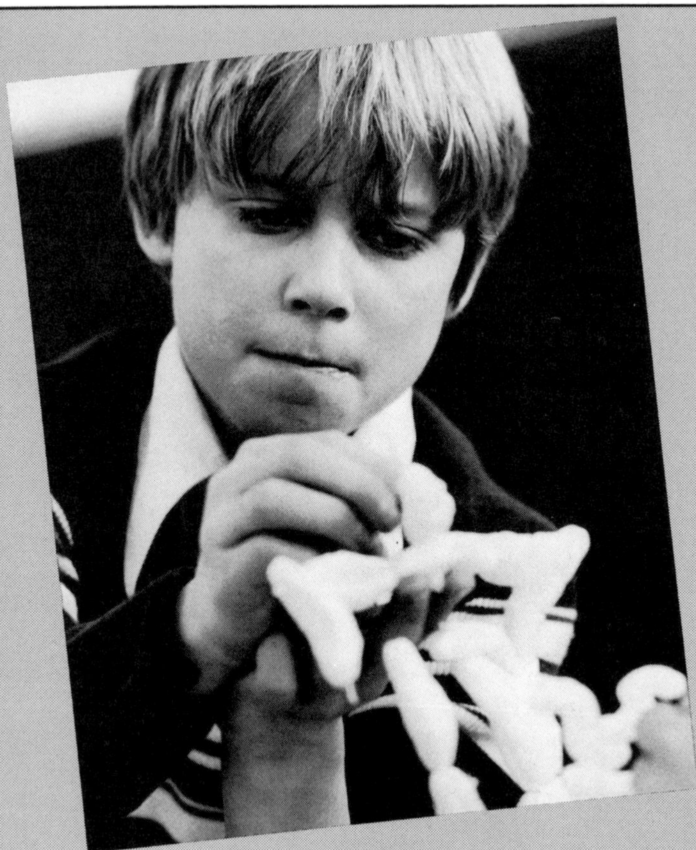

Projects, Competitions, and Family Activities

13 Hearing the Music of the Spheres: Informal Astronomy Education

Andrew Fraknoi
Astronomical Society of the Pacific
San Francisco, California

Whether it is the avid amateur astronomer scanning the skies from a dark country location, the youngster enjoying her first planetarium show, or the average worker stopping his or her morning routine to consider a point raised on the *Stardate* radio program—thousands of people, in all walks of life, learn about science and the scientific method through astronomy.

I can cite several reasons for astronomy's popularity, but those of us who are in the field may not be the best sources for objective explanations—we are, after all, clearly prejudiced. Still, I suggest that astronomy holds a special fascination because space represents the next physical frontier. And the discoveries of modern astronomy stretch our intellect and imaginations in ways that can outstrip the predictions of even the most "far out" science fiction.

Astronomy is also rare among the sciences because it provides an active and rewarding role for amateurs. With only modest equipment and a lot of patience and dedication, astronomy hobbyists discover new comets and exploding stars, monitor stars that vary in brightness in unpredictable ways, hunt for asteroids and fireballs, and spread their contagious enthusiasm for the skies among youngsters and adults in their community.

A variety of organizations and informal networks have grown up around astronomy, creating an institutional base for public education that simply does not exist for fields like particle physics or biochemistry. Although most teachers are probably casually familiar with these astronomical resource organizations, few are aware of the number or the types of services they provide.

Bringing the Skies Indoors: Planetaria

For most of us, the dark night sky that so captivated our ancestors and explains why astronomy was the first science can be enjoyed with only some effort. For city and suburban residents, the sky has been polluted by the lights of civilization (although many of us are not sure that fast-food restaurants and shopping malls necessarily qualify as civilization).

The second-best place to see and to become familiar with the stars is at a *planetarium*, where a specialized projector simulates the night sky on a dome. There are approximately 1,000 planetaria in the United States, ranging from small ones that accommodate about 20 people to giant facilities with 300 or more seats. Most planetaria are associated with a science or natural history museum, a college or university, a school district, or an individual school. Many have public shows and special programs or classes, although schedules and requirements vary tremendously. The International Planetarium Society can provide more information about building or operating a planetarium in your school district. To find the nearest planetarium, call the astronomy or physics department at a local college, your local science museum, or the science curriculum specialist in your district.

Radio and Television Programs

Most local public television stations have run and continue to run *Cosmos*, Carl Sagan's enormously popular and highly personal introduction to astronomy, and *Project: Universe*, a primer on modern astronomy. *NOVA*, the PBS science show, features two or three new astronomy programs each year, and many of these later become available for rental to school and community groups (but, alas, only at high prices).

Stardate is a daily series of brief astronomy reports on radio. Produced by the University of Texas MacDonald Observatory, the syndicated series has made millions of listeners more aware of phenomena in the sky and of modern astronomical research. Many radio talk shows feature astronomers as guests from time to time, although only Carl Sagan has broken the barrier by making regular television talk show appearances.

Astronomy Hotlines

Many planetaria, museums, and amateur astronomy clubs have telephone announcement lines with information about their local activities and sky events. Finding out about these can sometimes be difficult but is usually worth the effort. In addition, you can call two nationwide astronomy hotlines that will keep you up to date on what is happening in astronomy. These numbers are listed below.

On the west coast, the Astronomical Society of the Pacific (ASP) provides an astronomy news hotline that generally features a three-minute nontechnical report on a single development or discovery in astronomy. The number is (415) 337-1244. On the east coast, the staff of *Sky & Telescope* magazine offers *Skyline*, a recording that discusses several stories in less detail and also offers discussions of sky events for serious amateur astronomers. Their number is (617) 497-4168.

Both numbers operate 24 hours a day. Messages are changed once a week and cost nothing beyond the actual charge for the call.

Organizations for Armchair Astronomers

Most people begin exploring the universe from the comfort of their own easy chair—reading about astronomy, listening to the radio, or watching television. These "armchair astronomers" are distinguished from the *amateurs* who will add observing and maybe even telescope building to their activities.

Several nontechnical magazines are devoted to popular astronomy, including *Astronomy*, *Sky & Telescope*, and *Mercury*. The first two are available in many larger libraries, while the third comes with membership in the Astronomical Society of the Pacific. In addition, general science magazines like *Discover*, *National Geographic*, and *Science News* regularly cover astronomical ideas and discoveries. The AstroMedia Corporation publishes a colorful magazine for younger children called *Odyssey*, which contains a mix of education and entertainment.

The Astronomical Society of the Pacific is the oldest and most active national astronomy group in the country. Although the Society's name harks back to its origins on the Pacific Coast in 1889, today it has members in all 50 states and in more than 70 other countries. The ASP is unique among scientific societies in that it welcomes not only scientists and educators in the field, but everyone with an interest in the heavens. The Society publishes the popular *Mercury* magazine as well as a professional journal in astronomy, holds meetings and lecture programs, offers workshops on teaching astronomy, and gives five prestigious international awards. Through its *Selectory*, a catalog of educational materials, the ASP offers astronomical slides, posters, books, software, maps, and sky observing aids for teachers and the public.

The Society also offers a special free publication for teachers interested in using astronomy in the classroom. *The Universe in the Classroom* is a quarterly newsletter with information on new discoveries, practical classroom activities, and resources for teachers and students. Subscriptions to the newsletter are free, but requests must come on institutional stationery and must include the grade level the requestor teaches.

Another society especially for armchair astronomers is The Planetary Society, founded in the early 1980s by Carl Sagan and Bruce Murray and now the largest space-oriented group in the world. The Planetary Society publishes a colorful bimonthly magazine, *The Planetary Report*, holds conferences on topics relating to the exploration of the planets and the search for extraterrestrial life, and supports research in these areas. An important aspect of their work is to mobilize public and political support for the science and technology needed to understand more about our solar system. Their 100,000 or so members regularly write to Congress and work in their local communities to encourage the peaceful exploration of space.

Amateur Astronomy

Although many people are perfectly content to remain armchair astronomers, taking an occasional look at the dark night sky during a camping trip, others find they are bitten by the "observing bug" and want to observe celestial phenomena on a regular basis. (By the way, you do not need to rush out and buy a telescope if you are bitten. Instead, you can start observing with the naked eye or a simple pair of binoculars.)

Once someone makes a serious effort to become familiar with the constellations and begins to spend a number of evenings observing the sky, he or she has graduated into the category of amateur astronomer. It is here that an unparalleled network of local and national groups exists to help and to support those who want to learn more about the heavens.

At the local level, many cities and towns have an astronomy club, often a loosely organized group of devotees who meet once or twice a month, listen to guest speakers, hold "star parties" and telescope-making workshops, exchange information about the hobby of astronomy, publish an informal newsletter, and just enjoy one another's company.

Many amateur clubs are associated with a college, museum, or planetarium, but others exist only through members' efforts and meet in schools, churches, or members' homes. To find a club near you, ask the astronomy or physics department at the nearest college or a science museum.

Many, but by no means all, of the clubs are part of two larger federations or "umbrella groups" formed several decades ago. The Western Amateur Astronomers includes about 30 or so clubs in the western states, while the Astronomical League represents several hundred clubs throughout the rest of the country. These umbrella groups are run entirely by the efforts of volunteers, so if you write to them, remember to enclose a stamped, self-addressed envelope to help them with postage.

From the teacher's point of view, probably the most wonderful thing about local amateur clubs is that many of them have an ongoing program of bringing information to schools. With notice, a member will often come and talk to a class and even bring a telescope with a proper filter for daytime Sun observing. Some clubs also sponsor public observing sessions in the evening that students and their families can attend.

A much smaller subset of amateur astronomers will go on to perfect their skills sufficiently to actually participate in astronomical research. As mentioned earlier, amateurs have made and are continuing to make important contributions to astronomy in several areas. These include hunting for new comets, finding exploding stars in other galaxies (one amateur in Australia found 15 such exploding stars using only a very primitive telescope and a prodigious memory), and monitoring variable stars, which change their brightness in regular and sometimes irregular fashion. Although the details of these activities are beyond the scope of this chapter, it is worth pointing out that national groups bring together amateurs who practice them. For more information, write to such groups as the American Association of Variable Star Observers or the Association of Lunar and Planetary Observers (see resource list at the end of this chapter).

Astronomy on Computer

A growing area of informal science education in astronomy is in the field of home software, spurred by the enormous increase in the availability of reasonably priced home and school computers. A recent survey revealed more than 100 pieces of commercially available software for astronomy, ranging from simple calculational programs to complex sky simulation programs that convert your computer into a flat planetarium. (A reprint of the survey giving a full annotated list of these programs, with the addresses of all manufacturers and a bibliography, is available for $4 from the ASP.)

Debunking Pseudoscience

Alas, so much of informal education about astronomy these days turns out to be *miseducation.* Our students are constantly exposed to uncritical media coverage of "fiction science" such as astrology, UFOs as extraterrestrial spacecraft, and ancient astronauts.

In my own university-level teaching experience, a significant number of my students believe that the arrangement of stars and constellations at the moment of their birth can influence their destiny, that the Air Force is hiding the bodies of aliens who came here in a flying saucer, and that many of the great archeological monuments of the human species were built with the help of ancient astronauts (presumably, and insultingly, because our ancestors could not build them alone).

To help reporters, teachers, students, and the public evaluate the many claims of such "paranormal" phenomena, a new organization has been formed by concerned scientists, educators, and skeptics in all walks of life. Called by the rather awkward name the Committee for the Scientific Investigation of Claims of the Paranormal (CSICOP), the group includes such noted figures in science education as Carl Sagan, Isaac Asimov,

Stephen Jay Gould, and magician James Randi (who can spot a fraud a mile away). CSICOP publishes an excellent magazine called the *Skeptical Inquirer*, holds annual conferences, and has generated local groups in many cities in the United States and Canada.

Interdisciplinary Approaches to Astronomy

Because astronomy has inspired artists for millennia, it is not surprising that a great deal of informal learning in this field takes place through the humanities. There is, of course, the area of science fiction, which includes many good stories based on extrapolations of science and teaches a great deal about the universe to its many fans. But even in "serious" literature, astronomical ideas can inspire or challenge, as in John Updike's novel *Roger's Version* or the short story "Passions and Meditations," by Joyce Carol Oates.

In music, the many pieces inspired by astronomy include a jazz piece about supernovae, operas on the work of Kepler and Einstein, and even rock-and-roll songs about black holes. Many modern poets use images from astronomical research, and some, like Diane Ackerman and Diane Wakosky, have woven whole skeins of poems from astronomical themes.

A much more detailed exploration with many specific references to the interaction between astronomy and the humanities can be found in *Interdisciplinary Approaches to Astronomy*, a booklet available from the ASP for $4.

In its ability to inspire us, to stretch our imaginations, and to capture our fancy, astronomy is one of the best areas to introduce science to our students and to the wider public. As the poet Robinson Jeffers, whose brother was an astronomer, wrote earlier in this century in "Star Swirls,"

> There is nothing like astronomy to pull the stuff out of man.
> His stupid dreams and red-rooster importance: let him count the star swirls.

Resource Organizations for Informal Astronomy Education

American Association of Variable Star Observers, 25 Birch St., Cambridge, MA 02138

Association of Lunar and Planetary Observers, c/o Dr. John Westfall, P.O. Box 16131, San Francisco, CA 94116

Astromedia, c/o Kalmbach Publishing, 1027 N. 7th St., Milwaukee, WI 53233

Astronomical League, c/o Merry Wooten, Executive Secretary, 6235 Omie Circle, Pensacola, FL 32504

Astronomical Society of the Pacific, 390 Ashton Ave., San Francisco, CA 94112

Committee for the Scientific Investigation of Claims of the Paranormal, P.O. Box 229, Central Park Station, Buffalo, NY 14215

International Planetarium Society, Membership Information, c/o Mark Petersen, P.O. Box 3023, Boulder, CO 80307

Planetary Society, 65 North Catalina Ave., Pasadena, CA 91106

Sky Publishing, 49 Bay State Rd., Cambridge, MA 02238

Western Amateur Astronomers, c/o Stephen Edberg, Jet Propulsion Lab., 4800 Oak Grove Dr., Pasadena, CA 91103

14 Westinghouse Science Talent Search

Edward G. Sherburne, Jr.
Science Service
Washington, D.C.

Now in its 48th year, the Westinghouse Science Talent Search (STS) is unusual among scholarship competitions. Administered by Science Service and supported by Westinghouse, the competition places primary emphasis on the student's report of an independent research project in some area of science, engineering, or mathematics, and only secondary emphasis on academic achievement.

In short, the evaluation is based on the student's ability to "do" science in a way that is analogous to, though less sophisticated than, what a professional scientist does. To use a sports analogy, one does not test ability to play tennis by giving a paper-and-pencil test. One observes performance on the tennis court. We do not mean that academic ability is unimportant. But as Robert J. Sternberg (1985) comments in his book *Beyond IQ—A Triarchic Theory of Human Intelligence*, "Possession of knowledge does not guarantee the creative use of that knowledge." A high-quality independent research project could indicate creativity, as well as motivation, initiative, persistence, and other attributes that contribute to scientific performance.

To enter the Science Talent Search, a secondary school student does an independent research project and describes the research in a paper of about 1,000 words. The student also answers questions on a Personal Data Blank, which has open-ended questions designed to elicit evidence of the student's interest and creativity in science. These two items constitute an entry. The entry must arrive at Science Service by midnight, December 15. Any entries arriving late are automatically disqualified.

Scientists, engineers, and mathematicians from such institutions as Johns Hopkins, the National Institutes of Health, Princeton, and Berkeley work as judges. Entries are placed in categories by discipline and evaluated by at least two judges who are specialists in the relevant discipline. After evaluating all the entries, the judges rank the entries and designate the top 300 as honorable mentions.

The Admissions office of every four-year college in the country receives the names and addresses of the students, printed in a booklet that is sent with letters of recommendation. Because of the impressive undergraduate track record of previous honorable mention winners and the reputation of the STS program, the colleges approach many of the students with admission and scholarship offers.

The judges then study the honorable mentions, and from these, select 40 winners to compete for scholarships totaling $140,000—one $20,000 first, two $15,000 seconds, three $10,000 thirds, and four $7,500 fourths. The remaining 30 students receive scholarships of $1,000 each. STS awards scholarships without regard to financial need.

The 40 winners receive an all-expense-paid trip to Washington, where the judges interview the students. The students also visit scientific laboratories and talk with scientists about their work, see their members of Congress, attend the theatre (usually the Kennedy Center), and talk to the scientific public when their projects are exhibited in the Great Hall of the National Academy of Sciences. The scholarships are announced at a formal banquet on the last evening of the students' visit—the largest scientific dinner party of the year in Washington.

Many of the other entrants have yet another opportunity for recognition through the 35 State Science Talent Searches. Science Service duplicates the entries and sends them to the Directors of the State Science Talent Searches. The state STS's then conduct their own competitions, many of which have numerous awards, including scholarships.

Scientific Versus Academic Performance

Richard S. Mansfield and Thomas V. Busse (1981) comment on the "threshold effect" that has been suggested by some psychologists. They imply that a threshold of academic ability in a discipline is required, but beyond that level, additional academic achievement is not as important as other abilities such as creativity or motivation. In a sense, scientific performance guarantees a certain level of academic achievement because research can't be done without the necessary knowledge. On the other hand, what you know doesn't matter if you don't know how to use it. In a study of STS honorable mentions done by Harold A. Edgerton some years ago (1973), the scientific achievers (those with the best research papers) were not always those with the highest grades. The study looked at 300 honorable mention winners who were selected on the quality of their research papers, that is, on their scientific performance. These 300 were compared with another 300 from the total pool of entries selected for their academic achievement (grades, SAT scores, and class rank).

The study compared the two groups to see how many students were in both. Two-thirds of the students chosen on the basis of scientific performance would not have had high enough scores if academic achievement had been the criterion. And two-thirds of those chosen on the basis of academic achievement had such low ratings on their scientific achievement that they would have been excluded from the top 300 on the basis of scientific performance.

How Well Do the Winners Do?

If the selection process for the Science Talent Search is so different, how well do those selected do? One can look at this in two ways—in terms of college success and occupational success.

The results of a Science Service mail survey (64.9 percent response) of winners from 1942 through 1979 show that 99 percent of the winners have a B.S. or higher (Science Service, 1979). Even more impressive is that 70 percent of the group have a Ph.D., M.D., or both. Because undergraduate and graduate degrees only indicate college success, we need to look at post-college success (Hoyt, 1966). In science, post-college success is generally defined by how the scientific community recognizes the winners.

To date, five former Science Talent Search winners have won the Nobel Prize. That it probably takes 30 years after winning the STS to finish undergraduate and graduate school and do the necessary research is even more significant. And so, winners prior to 1960 are old enough and experienced enough to be eligible.

The National Academy of Sciences recognizes outstanding achievement in science through election to membership, which is considered the highest elective honor a scientist can receive in this country. To date, 28 former Science Talent Search winners have been elected.

For younger scientists, a fellowship award is an important recognition and often the precursor to later scientific achievement. One of the oldest and most prestigious programs is the Sloan Research Fellowships, aimed at stimulating fundamental research by young scientists of outstanding promise. To date, 47 former winners have received these Fellowships. Other major awards include two Fields Medals (the Nobel Prize of Mathematics) and eight MacArthur Fellowships.

This analysis of success cannot by any means be called "scientific," but it does suggest that, for whatever reasons, students selected by the Science Talent Search will succeed both in and after college.

The Importance of the Teacher

In one of the major contemporary studies of talent, Bloom (1985) concluded that the old expression "genius will out" does not hold. Whatever the individual's natural talents, strong support and influential teachers are essential in helping a student reach high levels of achievement.

One might correlate winning the Science Talent Search with attending a good school. This is not completely true. The schools that produce winners and honors are usually good, but many good schools do not produce any winners or honorable mentions. Many do not even have entries. Some schools produce winners for a while and then, for no apparent reason, stop. We found that usually a particular teacher left that school. If the teacher moved to a school that had no winners or honors, the school began producing winners. Nothing changed but the teacher's location.

Thus, the overall quality of a school seems unimportant. Instead, the school must have one or more teachers who want to develop student research, and more specifically, who want to encourage students to enter the Science Talent Search.

What characteristics do these teachers have? A study of the Science Talent Search (Campbell, 1983) found that they have exceptional knowledge of and enthusiasm for the subjects they teach, and they can communicate those qualities to their students. They are interested not only in the subject in an academic sense, but also as an area of ongoing research to develop new knowledge. They are hardworking and often put in long hours at night and on weekends. They have close relationships with the students, and working with the students on the projects deepens those relationships.

In the Personal Data Blank students name one person who influenced their scientific interest the most. Students mention teachers or school staff most frequently, scientists second, and family members third.

References

Bloom, Benjamin S. (Ed.). (1985). *Developing talent in young people.* New York: Ballentine Books. (Paperback).

Campbell, James Reed. (1983). *Horizontal enrichment for precocious high school science students.* Paper presented at the Annual Meeting of the National Association for Research in Science Teaching, April 7, 1983, Dallas, Texas.

Edgerton, Harold A. (1973). *Identifying high school seniors talented in science.* Washington, DC: Science Service.

Hoyt, Donald P. (1966). College grades and adult accomplishment—A review of research. *The Educational Record.*

Mansfield, Richard S., & Busse, Thomas V. (1981). *The psychology of creativity and discovery.* Chicago: Nelson-Hall.

Science Service. (1979). *Survey of Science Talent Search Winners.* Washington, DC: Science Service.

Sternberg, Robert J. (1985). *Beyond IQ—A triarchic theory of human intelligence.* New York: Cambridge University Press. (Paperback).

Helpful Hints

■ Research:

Identify promising students in the ninth or tenth grades and have them learn about and do research.

Start students early on their research projects. Some projects represent months or even years of work.

Do not let students wait until the last minute to write their papers, which also takes time.

Refer students to the *Directory of Student Science Training Programs for High-Ability Precollege Students* published by Science Service if they are looking for summer research opportunities.

■ Entries:

Mail entries early. Remember, the deadline is midnight, December 15, for receipt of the entry, not the postmark. If you're near the deadline, use an overnight delivery service.

Requests for entry materials must be from a teacher, guidance counselor, or principal on school letterhead, and materials must go to the school. Parents and students may not request materials.

Have students fill in the Personal Data Blank first. The teacher may then add comments and provide transcripts of tests scores and grades.

Encourage students to be concise and to the point. The quality of the writing, not the quantity, counts.

■ Rules:

Avoid library research.

Read the rules carefully and note that they may change in some way from year to year.

■ Recognition:

Recognize participants and especially honorable mentions or winners through a school assembly, the school newspaper, local newspaper, etc.

Last, write Science Talent Search, 1719 N St. NW, Washington, DC 20036 if you have questions.

15 Training Young Researchers

Sol Lander
New York City Board of Examiners
Brooklyn, N.Y.

Each year New York City students competing in the Westinghouse Science Talent Search consistently shine in both the number of honors and winners they achieve and in the quality of their research. In the 1979–80 competition, more than 60 New York City students were recognized; the following year, 82 students were winners! In a more recent competition, 12 New York City students achieved honors out of the 40 nationwide finalists, with four of them receiving honors as the first, sixth, ninth, and tenth top prizewinners.

What is the key to this consistent success? Although official training is not provided, students are encouraged at an early age to participate in school, district, and citywide science fairs. Let's look at the step-by-step process for training young people to compete in science research projects.

Steps to Training Young Researchers

Search and Identification. Assistant principals

visit intermediate and junior high schools in their neighborhoods, talk to graduating youngsters and Special Progress classes, and explain that strong science programs are available in the principal's particular high school. (Students who are two years above level in reading and mathematics are eligible for seventh- and ninth-grade Special Progress classes. An expanded, more demanding curriculum and opportunities for individualized research in social science and science are features of such classes.) This step is greatly simplified in small school districts. Principals examine records of science-oriented students, noting special interest and achievement in science and mathematics as well as information on independent research such as entry in the annual citywide science fairs. Assistant principals perform much of this investigative work in their free time, often on the weekends. In New York City these fairs are conducted under the supervision of the American Institute of Science and Technology and the Science Unit of the Division of Curriculum and Instruction.

Personal Interview. A student interested in entering the competition receives a personal interview at the high school. The assistant principal tries to determine the student's maturity, enthusiasm, and dedication, and spells out the requirements clearly: reading science journals and magazines, doing a literature search, working additional hours, and having diligence and determination. These preliminary steps are vital for the involvement of qualified youngsters.

Guidance. The principal helps the student explore his or her area of interest. In this step, student and principal determine the practicality of problems to be researched, availability of literature sources, and the time the investigation will require. Often, classroom discussion motivates a student to decide to pursue a particular project. Television programs and magazine articles are additional sources of motivation. Exposure to a variety of science topics is the key. The cooperating teacher's role (the next step) becomes important at this point; overambitious students must be careful not to select an experimental problem that is over their heads.

Selecting a Capable Cooperating Teacher. Guidance, direction, counsel, and research experience are some of the main qualities principals consider when selecting a cooperating teacher. The teacher must be available to the student, and this may mean that a teacher normally required to teach five classes per day will have one class or an administrative assignment removed from his or her program.

Consultation. After the student has some idea of the areas to be researched, he or she consults with the teacher. The teacher and student discuss

- time frame for the project
- basic elements of a good project
- project notebook
- use of graphs and statistics
- data interpretation
- writing the report, including the bibliography

Program Planning. Aside from the regular Regents science and mathematics courses (applicable in New York State), students must take courses in advanced placement and laboratory techniques, and they must participate in summer institutes such as those sponsored by the NSF or offered at local colleges and research institutes. They also must learn how to use community facilities such as the New York Academy of Science and the Sloane-Kettering Institute. Students are encouraged to participate in project competitions, including the City-Wide Science Fair, National Energy Foundation (SEER Energy Program), Junior Science and Humanities Symposium, Westinghouse Science Talent Search, and the Otto Burgdorf Science Conference.

Recent developments in the Otto Burgdorf Science Conference make the conference especially valuable training for students. University facilities are made available to the prizewinning students to explain their projects and their investigatory techniques to other high school students invited to attend. A question-and-answer period follows on aspects of the particular piece of research and general aspects of reports, data, graphing, and presentation. This procedure serves as a format for peer exchange and oral presentation, helping students to express themselves and clarify the tools of science.

Supervising teachers also conduct sessions at this conference with the students and with one another, where they outline techniques in project report progress and self-evaluation. The student, therefore, has the opportunity for self-evaluation, and this evaluation is reviewed with the teacher.

Formal Notification of Parents. Parents also may become involved. The school sends parents a letter advising them of their child's interest in an accelerated science program and explaining the full complement of available courses in tenth, eleventh, and twelfth grades. The school informs parents of afterschool participation in science competitions, the Columbia University Saturday Science Program, and the National Science Foundation Summer Programs. The parent is invited to sign the letter, return it to the school, and call if any questions remain unanswered.

Coordinating Meeting of Supervisor, Teacher, and Laboratory Specialist. The student meets with supervisor, teacher, and laboratory specialist to discuss time, materials, and practicality of the project. The student must show that he or she has done library research in this area. Participants in the meeting discuss the cost to the school and to the students for materials. The student and the supervising teacher arrange days and hours of consultations, and the wonderful world of research welcomes another young investigator!

General Outline and Breakdown of Time Slots. Library research requires four weeks, equipment description two weeks, the experimental phase three weeks, data collection three weeks, data interpretation and analysis three weeks, and report writing three weeks. The total? Four and one-half to five months for the entire procedure. Students find this breakdown extremely helpful as both a daily and an overall guide for each phase; they know that procrastination and meandering may lead to incomplete projects. The entire process may not necessarily be completed in the five-month period but can take up to two years from conception to final presentation, depending upon the student's nature or method of working.

Westinghouse Reception. The final aspect of the competition is the Westinghouse reception held in a New York City hotel. The Board of Education of New York City goes to great lengths to ensure a planned, effective program. We thank the Westinghouse Electric Corporation, the Westinghouse Educational Foundation, and Science Service for initiating and maintaining this prestigious event for more than 40 years. As long as the event endures, New York City will participate to the fullest extent possible and continue to shine in educational and scientific achievement.

How wonderful it is to behold the exploration of the scientific world through the eyes of teenagers! Challenged by competition, motivated by their own interests, and eager to work diligently toward recognition, these young people open their minds to the possibilities of the future for themselves and for the world.

16 Family Math: Making the Home an Environment for Problem Solving

Kathleen Devaney
Berkeley, California

At 6:30 p.m. on a late-October Thursday, the first students arrive in the library of Bayside Elementary School, a K–5 building in a middle-income family neighborhood. A tall man wearing casual clothes and running shoes, and two boys, one about 8 years old and the other about 10, stop at the sign-in table just inside the library door and print their names—"Steve," "Rob," and "Tim," on stick-on name tags. Steve and Tim attach theirs to their sweaters; Rob pastes his on his Baltimore Orioles baseball cap, which he wears backwards.

At the sign-in table, Tim, the older boy, picks up a glass peanut butter jar filled with raw kidney beans and bearing the label: "Estimate how many beans are in the jar." Tim turns the jar around. The three confer but do not agree, and write three answers on separate slips of paper and place them in the answer basket.

Susan Sherman, one of the two co-teachers of this class and a third-grade teacher in this school, greets them by name and points them toward the sign-in graph posted on a corkboard. The graph is a Venn diagram of three circles, "I am a sister," "I am a brother," and "I like to play Balloon Ride" (an activity they learned at last week's class). All ponder the graph, and after some pointing and talking among themselves, Steve, Rob, and Tim write their names in the "brother"/"balloon ride" intersection on the graph.

On the opposite side of the room, Holly Benson, the other co-teacher, has just finished setting up learning center instructions for self-paced activities. She suggests to Steve that he and his sons begin to work on one activity and then teach it to the next family that's arriving—a young woman holding a baby in a carrying seat, and a girl about age eight. Before they finish signing in, the father arrives and takes the baby, while mother and daughter read the Venn diagram and print "Sarah" and "Rebecca" in the "sister" circle.

This is a hypothetical, but typical, Family Math class. Each family does hands-on activities together using inexpensive household materials. They also get handout instructions for doing the same and other activities at home. All the activities are designed to be done over and over without boredom. Thus parents learn, through their own and their children's accomplishments, that math skill is not a gift from God but a gain from repeated, concentrated, inventive, frequently cooperative work and play. They come to understand that math ability is not rare; it need not be painful to achieve. Anyone can do math, and it can be fun.

Family Math has three aims: to stop mothers and fathers from passing their null or negative attitudes about mathematics on to children; to help parents create an environment—at home with family activities such as household jobs, shopping, trips, and games—that familiarizes children with the broad scope of math; and to persuade parents and children to approach math as problem solvers. Parents learn they can stimulate their children's mathematical and scientific thinking in the home just as naturally and enjoyably as they nourish their children's literacy by reading to them.

The Beginnings of Family Math

Started by the University of California's Lawrence Hall of Science in Berkeley in 1982, Family Math began as a response to black mothers in a low-income neighborhood of Richmond, California, who asked Clyde Wallace, the assistant principal at their children's school, to teach them the math they needed to help their kids at home. Wallace mentioned their request to teachers of the EQUALS course he was taking at the Lawrence Hall of Science. (Through inservice programs, EQUALS helps teachers retain students who otherwise would not succeed in math—especially girls, blacks, and Hispanics—by persuading them that they can and should continue taking math even when it is no longer required.) Taking Wallace's suggestion, Virginia Thompson, an EQUALS staff member who had been a math resource teacher, designed a course that parents could attend with their children, learning math through activities that could be repeated with enjoyment many times at home. Since 1982 when Thompson piloted the first class in Richmond, Family Math has spread across the United States. More than 15,000 participants—children and their parents—have taken the class.

Basic Principles of Family Math

Family Math teaches the relationship between mathematics and the natural world and work life. Parents learn that perseverance in math is vital to *all* children's school preparation for higher education and for most rewarding careers. They learn to emphasize the usefulness of mathematics in problem solving. Parents realize that math means more than just arithmetic and algebra, because our science-oriented society requires abilities to measure, estimate, visualize spatial relationships (expressed through geometry), interpret data using probability and statistics, reason, and solve problems logically. Family Math cultivates this *problem solving*—using the mind accurately, analytically, systematically, investigatively.

The program originators have identified the basic principles of Family Math for adaptation outside of school. These principles also can be applied to science subject matter.

Focus on problem solving. Teach parents and children to *think* about a problem by looking for patterns, drawing a picture, working backwards, working with a partner, or eliminating possibilities. "Having a supply of strategies allows a choice of ways to start looking at a problem, relieving the frustration of not knowing how or where to begin," write program directors Nancy Kreinberg and Virginia Thompson (1986). "The more strategies you have, the more confident you become, the more willing you are to tackle new problems, and the better problem solver you become."

Teach with "hands-on" materials. Household objects like beans, blocks, string, pennies, playing cards, and toothpicks help children and adults understand numbers, shapes, quantities, space, pattern, and relationships. "Traditionally, these materials are used mainly in early elementary years, and paper-and-pencil mathematics becomes the rule after second or third grade," Kreinberg and Thompson write. "This is unfortunate, since much of mathematics can best be explained and understood using the tools of manipulative materials and models; and, in fact, many research and applied mathematicians do just that."

Arrange the class environment and the schedule for serious but nonstressful efforts by parents and their children. "We provide a supportive environment in which parents and children feel comfortable doing mathematics—an environment that is at once nonthreatening and encourages risk taking. . . . We present familiar math topics at first and gradually introduce ways of doing math problems that are unfamiliar, so that curiosity is piqued and motivation to solve the problems encouraged." In Family Math, there are no grades, credits, or judgments; parents consistently observe and experience adults working with children in nonjudgmental ways.

Demonstrate and explain the scope and sequence of mathematics, the specific topics being taught at the children's grade level, and how topics relate to each other. Present mathematics content in a way that explains why children are taught particular concepts at certain ages, how these concepts are interrelated, and why learning math is important.

Teach activities parents can do at home with their children. "Parents learn activities they can use at home to reinforce the concepts, and that are interesting enough for repeated use without falling into a drill-and-practice mode."

Introduce the link to careers and the future. "To ensure that the reason for studying mathematics is made explicit, men and women from the community, working in math-based occupations, come to Family Math classes to talk about how math is used in their jobs and the jobs of people with whom they work."

Model a teaching style that generates enthusiasm and persistence. "We provide early success so that learners will want to continue; encouragement to move at a pace that is comfortable for [each] learner; and an ambience of informality, vivacity, and variety that stimulates and sustains interest." In the atmosphere of cooperation, thinking out loud together, and friendly challenge, people dare to try something new.

Family Math can be taught in classrooms, the library or multipurpose room of a school, in a church fellowship hall, a public library, a science museum, a community center, or a boys' or girls' clubhouse. It can be taught by teachers, teachers' aides, parents, youth workers, retired persons, or older students. It can be taught as one session—for instance, one whole Saturday—or as six weekly sessions, each approximately two hours, held in the early evening. Classes can be as large as 200 or as small as five.

Every Family Math class is different because each instructor chooses from a collection of more than 100 active learning lessons explained and pictured in the Family Math book (published by and available from the Lawrence Hall of Science, University of California, Berkeley, CA 94720), plus the teacher's own old-favorite or newfound activities. Usually, Family Math leaders target classes to two or three grade levels of children, for instance, K–2, 3–4, 5–6, 7–8.

An Imaginary Typical Class, Continued . . .

Back at the activity stations, Steve, the father, extends a roll of adding machine tape from Rob's shoe sole to the top of his head and cuts off a strip, while Tim does this for himself. Then Steve holds Rob's strip against his sideways-stretched arms and marks on the strip the length of the boy's reach. The second family arrives, and Sarah, the mother, helps Steve take his own measurements, while Rebecca is measured by her father. Tim copies from a sample card a chart on which to enter his family's findings: Under the heading "Short rectangle (the tape is shorter than your reach)," he prints Rob's name; under "Tall rectangle (longer than your reach)," his father's name; and under "Perfect Square (about the same length)," his own name. Rob sticks his arms out so that his elbows make right angles, tramps a square path in front of the table, and chants, "Tim's a square—a perfect square!"

Now families are gathering around the Venn diagram, as the rest of the class, about 15 families in all, arrive just before and after the 6:45 p.m. starting time. About one-third of the family groups include both mother and father with one or more kids. Once signed in, families move to the activity stations, all dealing with measurement. Besides "Perfect People," there are activities about volume (involving different sized jars and plastic cups to be filled with beans); about area and perimeter (an old *TIME* magazine, a blue-and-brown poster for the upcoming school science fair, and a cracked Barbra Streisand record, all to be covered with white beans or one-inch squares); and about mass (using small lumps of clay and a balance made from an old foot ruler, string, paper clips, and paper cups).

The room is buzzing with chatter. Families who finish all the activities sit at tables around the room and follow instructions on a handout sheet entitled "Value of Words." Each letter of the alphabet is assigned a monetary value, starting with one dollar for "a." Each member of the family is to calculate mentally, if possible, the value of his or her name; then the family finds the sum of the values of all their names and looks for another family with the same name value.

A little after 7:00 p.m., Holly calls everyone to be seated around the library tables. "Let's talk about what you just did," she begins. "What kinds of measurement were you doing? What activity measured how long something was, or how high?" Several children raise their hands and shout at the same time, "Perfect People."

"Were there any activities where you found how many beans or squares it took to cover something up?" Acknowledging the children's answers, she continues, "That's called area." She similarly identifies the activities in volume and mass.

"When you covered the record and the poster with the beans, we might say that you were measuring with *nonstandard* units. They work fine to help you compare the area of the record, which is a circle, with the area of the poster, which is a rectangle." Holly manages to address both the children and the parents. "In school, students learn first to *compare* measurements, then how to *order* measurements from small to big, and then they start using *nonstandard* units, such as string or toothpicks, beans, or squares; and finally they go on to *metric* and *English standard* measurement."

A mother asks, "Why do our kids have to learn about the metric system? I don't use it, and I think it's fading out—like at the gas stations—because people don't understand it."

Holly replies, "Yes, of course students are learning both English and metric, because they'll have to use both." She asks for examples of work in which knowing metric is required. Parents volunteer answers of car repair, manufacturing products for export, engineering, medicine, and sewing with European dress patterns.

Susan takes this opportunity to pass out a handout from the Family Math book. It defines standard and metric units of measurement for length, area, capacity/volume, weight/mass, and temperature. Holly calls attention to the comparisons: "It's not hard; just remember that a liter is a little more than a quart; a meter is a little more than a yard. It is important for children to be familiar with both measurement systems, but they probably won't have to figure out exact conversions from one to the other until they get to high school physics."

Concluding this five-minute overview of measurement, Holly next asks the group to report on the activities they did at home the past week using handouts from the previous class. An eight-year-old girl says she taught her friends to play "Balloon Ride," a variation of a Chinese game called "Nim." Toothpicks represent ropes holding a hot-air balloon to the ground. It develops intuitive understanding of subtraction and logical thinking. A boy says his whole family played "Double Digit," a dice game relying on chance—probability—and skill in estimation, and he won. A mother says she and her kids did the volume ordering activity—discovering which of her pots and pans hold the most water— and she was the one who spilled water all over the kitchen floor.

At 7:15 Holly and Susan split the group, keeping families together. Susan teaches an activity called "Create a Puzzle." As she passes out scissors and square pieces of cardboard, she explains that this activity develops understanding of the attributes of geometric shapes. Each adult or child makes three straight cuts in his or her cardboard and then figures out how to put the four pieces back together into the original shape. Then they give their puzzle to a partner to solve. "The more chances people get to handle real materials," Susan explains, "the more they gradually develop the ability to visualize the relationship of objects in space.

"This is a very important skill—for instance, in reading and sketching maps, giving and following place directions, understanding diagrams and illustrations for putting toys or furniture together. If you practice these skills at home, you will strengthen children's confidence when they have to tackle geometry in high school. Take a look at a geometry book. Three-dimensional figures start on page three! Unless students have visualized figures in space when they're much younger, that can be hard. So children need lots of cutting, and putting back together, and pasting, and putting things on top of other things. If you've cut apart an isosceles triangle and put the pieces together to make a rectangle, you are more likely to understand eventually why the formula for the area of a triangle is half of its base times its height."

On the other side of the library, Holly teaches "Two-Dimensional Nim," a game played by pairs taking turns putting their respective markers (bottle caps and beans) on a 3x6 rectangular grid. Holly explains that this is a reasoning game. "Play this often, and you'll get better strategies for winning—and for solving all kinds of problems," she promises. "Strategy means saying 'What if . . .?' It means being able to experiment and take risks. If you think you have a strategy that can win all the time, I'll play it with you; if you win, I'll change the rules to make it harder."

At 7:30, the teachers put out juice and coffee they have provided and cookies the parents have brought. For 10 minutes adults chat, children play, and some families finish measurement activities they didn't do at the start of the class. After the break, the two groups switch activities. When they finish, the teachers separate parents and children, and Susan teaches the children how to play "Balloon Ride" on the calculator, two kids to a calculator.

Holly takes the parents into an alcove, where she talks with them in detail about how tonight's activities fit into the school's math curriculum. One of the prime motivations for teaching Family Math is to counteract parents' belief that math is mainly numbers—arithmetic. So bit by bit throughout the series, they explain the rationale behind the state's new mathematics framework and how each topic will prepare children for higher mathematics and science study. She invites parents' questions and encourages their examples of things they can do at home to enrich children's understanding and enjoyment of math.

Most of this 15 minutes is dialogue. Some parents express frustration about their own poor math instruction and surprise at their new discovery that they can do this work without struggle or boredom. Holly reassures the parents that children will learn if they fully understand underlying concepts before they encounter new work, and if they are not made to fear failure and to hate the drudgery of conventional drill.

Holly is trying to influence the parents who push their kids too hard and the parents who do not help their kids at all. For instance, Steve seems skeptical about the profundity of "Double Digit," which he has played with Rob and Tim the previous week. "I don't see them doing all this deep thinking while they're playing."

Sarah interjects, "Oh, I think they just play! But after a while they begin to see a pattern." "Right!", Holly adds. "All these activities are designed to be repeated. As they play "Double Digit," they are bound to develop intuition about probability. Later in school, this intuition will be made explicit to them. Everyone accepts that parents can develop their children's reading enjoyment by reading stories to them. In the same way, you can develop math literacy and enjoyment if you play these games and puzzles and solve everyday problems with your kids, and identify the mathematics that's everywhere in your home."

The class reassembles for closing activities. Susan announces that Floranetta, a 10-year-old, has won the beans estimation with her guess of 637. Floranetta gets a big hand and explains how she came within three of the right number. "I looked through the bottom, and I guessed there were about 30 across there; and I turned it on the side, and I counted 20—well not exactly—*about* 20 going up to the top; and so I multiplied and I got 600; but then I thought about the ones that were falling through the cracks in the rows, and I just guessed, like, 637."

"Don't be afraid to guess!", Susan exhorts. "That's what estimation is—careful, informed, practiced guessing."

Holly passes out this week's homework: Find out all your friends' favorite ice cream flavor, and make a pie chart to display your results.

The final group activity is a human pie chart. "Suppose there were only three flavors of ice cream: vanilla, strawberry, and chocolate," Susan says, standing in the middle of the whole class, who are now on their feet. "All who love strawberry ice cream stand in a nice arc here to my right." Holly helps arrange the strawberry aficionados while Susan continues: "Now all the chocolate-lovers make your arc continue around the circle as far as you reach." Finally, they assemble the vanilla-lovers in an arc, completing the circle. Holly pastes three lines of masking tape from the center out to the circumference, making boundaries between the flavor arcs.

"We've made our own ice cream graph," Susan says. "Which flavor has the most people?" The children shout, "Chocolate! Even if we didn't count, we'd be able to see that, because chocolate is the biggest piece in our pie chart." Susan says, "Now can you see the way you can show what you find out about your ice cream-loving friends? You can draw a graph on paper like the circle we've made here. A graph is a *picture* of the information you collect."

Before the class closes, Holly announces that next week, "role models," a carpenter and a mechanical engineer, will describe how math is essential in their work and how they decided on these careers. Holly invites the children to bring their older brothers and sisters to the class. Holly and Susan have recruited a woman carpenter and a black engineer to prove that white males are not the only students who can aspire to math-based careers.

As the families leave, some remain to finish the measurement activities as Susan and Holly begin packing their materials. They are tired at the end of a day that started early—this class on top of a full day's classroom teaching. But they have found that this teaching creates rather than drains energy.

Why Teachers Volunteer for Family Math

Family Math gives teachers the exciting opportunity to work as a team, as well as the new challenge of teaching adults and children together. Family Math allows teachers to give information directly to parents and to stimulate their interest in improving the math curriculum to prepare *all* students for high school math and science.

Family Math is a way to explain to parents the state's or district's math curriculum and rationale. Through active learning and by watching their children, parents become convinced that the new methods work better than the way they were taught.

For teachers, the classes also are a source of new learning. Observing the interaction of children with their parents gives insight into students, and the activities give teachers ideas for improving their own classroom instruction. Holly and Susan can't help but feel energized by such rapid attitude changes and accomplishment in both children and adults. By giving children a comfortable and relaxed work and communication experience with their parents, they contribute, at least for the present, to the emotional health of these children. They know that for some, Family Math is a powerful contrast to family arguments over schoolwork—nagging by the parent, resistance by the child. "Homework without frustration and tears" is the way one parent describes Family Math.

Interviews with teachers who have taught several classes of Family Math reveal all these as reasons for taking on another teaching responsibility, frequently as volunteers. Some Family Math teachers who teach classes in which their own students participate cite evidence that some children's test scores improve after Family Math. But the program is too brief in duration (10 or 12 hours at maximum) and too young in practice (six years) for teachers to expect that it will raise students' achievement. What Family Math teachers are after is something more complex, subtle, and far-reaching: gaining the partnership of parents in their children's learning.

The Family Together, *Talking* Problem Solving

Disappointment and frustration about parents who keep at arm's length from their children's schooling is one of the biggest detractions in teaching, according to Stanford University researchers who interviewed representative teachers about their sense of satisfaction and effectiveness (Lareau, 1986). Family Math not only involves parents in helping their children with homework, but it also stimulates the playing, talking, and thinking together that teachers see as indispensable in emotionally caring for children.

Family Math provides an informal setting where teachers and parents can get to know each other over time and where teachers can become more sensitive to parents' needs, as well as explain the school's rules and needs. Thus Family Math appeals not just to "star" teachers and to math specialists, but also to "ordinary" teachers who want to gain parents' participation with children in learning.

Still, most teachers who undertake Family Math teaching come to it convinced that what they need from parents is not only enforcement of homework assignments and reinforcement for school rules, but also a new spirit or *ethos* about mathematics, and more *expressions* of mathematics in everyday life at home.

"I introduced to them that their home is a lab," says Carolyn Gray, a math resource teacher in San Bruno, California. She has taught several series of classes at her school, which she describes as an ethnic rainbow. "I told them they can do math in the kitchen, or at the beach. A favorite activity was cutting salami and fruit or vegetables. I told them, 'You know what shape the tomato will be when you cut it, but your kids don't.' Every time we were doing an activity, I was stressing that it was open-ended and unfinished; they could go home and continue with it.

"Some adults don't want to do the activities because they are threatened that the kids will get the answers before they do. They think it's a competition. I told them, 'I don't want the answers. I want to know *your mind*—how you get to those answers.' "

And Terry Juhl, who has taught several series of Family Math in both rural and urban schools in the Sacramento area, says, "The most important thing is the family together *talking* math, *talking* problem solving. That is very, very valuable. They are both learning together—this is rare."

Teachers like Gray and Juhl teach math to low-income and immigrant parents, some of them non-English-speaking. Bill Ruano, a fourth-grade teacher, holds his classes in his school in Walnut Creek, an upper-middle-class suburb of San Francisco. Ruano finds the same values in Family Math that Gray and Juhl speak of. He cites as common this comment from a single parent: "Family Math was the best quality time Wendy and I have spent together in a long time."

"In Family Math," Ruano says, "there's sharing between parents and kids. Children have an equal chance to share their knowledge, to win at games. That's unusual. Children are usually the little people, and parents have all the power."

By adolescent years, the power balance may begin to shift, and then Family Math can be valuable in setting an example of mutual cooperation and positive communication between teenagers and their parents. Kathy McKeown, an eighth-grade coordinator at a Berkeley junior high, organized a Family Math class in geometry, which students and their parents were motivated to take because of the tough course waiting for them at Berkeley High School. McKeown observed, "Family Math bridges the gap. It opens communication between parents and kids on things they ordinarily don't talk about."

This potential in Family Math to expand the content of communication as well as balance the power within the family has implications not just for emotional well-being but for students' intellectual development as well. Children spend worrisome amounts of time in entertainment pursuits with peers rather than in interaction with adults. (A Foundation for Child Development study found that women who work in the home spend an average 27 minutes a day in education-related activities—playing, talking, reading, teaching—while women who work outside the home spend only 11 minutes. Fathers spend about nine minutes a day in such activities with their children.)

"There is a very important need for interchange and communication between generations," says McKeown. "Adolescents need communication with their parents on *content* instead of struggles over authority. In Family Math you're building a different kind of communication, allowing kids to communicate about the content of the world."

Organizing Family Math Courses

So far, classroom teachers have initiated and taught Family Math for their own students. They organize the course after taking a two-day workshop in which they do a dozen or more of the activities in the Family Math book, receive the book as a teaching resource, and get tips on how to schedule, publicize, recruit for, program, and conduct a series of classes tailored for their own community.

Teachers find they must spend considerable time planning, but once a class is organized, it is self-guiding. Family Math teachers do very little lecturing, which would turn off children and adults.

Workshops for Family Math leaders have been taught at the Lawrence Hall of Science in Berkeley, Los Angeles, Orange County, Santa Barbara, San Diego County, Portland, Minneapolis and St. Paul,

Charlotte (North Carolina), Indianapolis, Phoenix, Nashville, Massachusetts, Vermont, Idaho, and New Jersey. The program has spread to Australia, New Zealand, Canada, Sweden, and Puerto Rico. The Family Math book is also available in Spanish and Swedish.

The program is popular with middle-class parents. In some suburban communities, Family Math becomes the thing to do, like Little League. But reaching parents who aren't involved with their children's learning—who leave it to the school and the professionals—is more difficult. In such communities, teachers need the assistance of classroom aides, bilingual teachers, PTA members and other parents, community liaison workers, counselors, and the school principal.

Increasingly, schools pay to send their teachers to Family Math workshops. Administrators value the program as a way to explain the enriched mathematics curriculum to parents, improve schools' public relations with parents, and above all, lure parents who would not otherwise respond to the schools. Some schools serving children of immigrant parents find Family Math is a way to explain they are needed as teachers at home, since math is not dependent on parents' knowledge of English.

In some instances, community agencies may be more effective than schools in organizing Family Math programs. In Indianapolis, the Urban League conducts classes to train a cadre of parents who will teach Family Math in their neighborhood.

In Phoenix, a multiservice agency for the Hispanic community, Valle del Sol, sponsors classes in Spanish and English.

In Oregon, Family Math is spreading through a statewide network of American Indian educators who teach Indian parents to conduct classes in cities and on reservations.

In Washington, D.C., the National Urban Coalition trains parents to teach Family Math in black churches.

In San Ysidro, California, a Hispanic community organization called Alba sponsors Family Math among Spanish-speaking parents.

The NSF funds these projects.

Mary Jo Cittadino, a parent from Albany, California, who took Family Math to help her first-grade daughter, has become a Family Math instructor in her school district and at the Lawrence Hall of Science. She credits Family Math with "revolutionizing" her own education—making her "math competent."

"I'm a real proponent of the Family Math style—active learning," Cittadino says. "It gives you willingness to [take] risks. The consequences aren't that terrible. You learn from your mistakes. Often, mistakes are more important than getting the right answer. This carries over into other aspects of life.

"I've learned I'm capable. I knew I was quick to understand certain things, but where I excelled was where I put my time. Now I know that everybody can do math; we just do it at different paces. I've learned there are steps to doing things mathematical that can be learned, that don't have to be intuitive. If we are trained, we can develop understanding and feeling. Math is a whole new way of perceiving the world. Nobody should be cut off from it."

References

Kreinberg, Nancy, & Thompson, Virginia. (1986). 1983–1986 Report of Activities to the Fund for the Improvement of Post-Secondary Education. Washington, DC: U.S. Department of Education.

Lareau, Annette. (1986). *Perspectives on parents: A view from the classroom.* Stanford, CA: School of Education Policy Institute, Stanford University.

Helpful Hints

- Let your children know that you believe they can succeed. Let them see you enjoying the activities, liking mathematics. Children tend to emulate their parents, and if a parent says, "You know, this is really interesting," that becomes the child's model.

- Be ready to talk with your children about math and to listen to what they are saying. Even when you yourself don't know how to solve a problem, asking a child to explain the meaning of each part of the problem will probably be enough to find a strategy.

- Be more concerned with the *processes* of doing mathematics than with getting a correct answer. The answer to any particular problem has very little importance, but knowing *how* to find the answer is a lifetime skill.

- Try not to tell children *how* to solve the problem. Once they have been told how to do it, thinking usually stops. It's better to ask them questions about the problem and help them find their own methods of working it through.

- Practice estimation with your children whenever possible. Estimation helps the thinking about a problem that precedes the doing, and is one of the most useful and "sense-making" tools available.

- Provide a special place for study, allowing your child to help you gear the study environment to his or her learning style. Some kids really do work better sprawled on the floor or bed, or with a musical background. There are no hard and fast rules.

- Encourage group study. Open your home to informal study groups. Promote outside formal study groups related perhaps to scouts, church, or school organizations. This will be especially important as your children grow older.

- Don't expect that all homework will be easy for your child or be disappointed that it seems difficult. Never indicate that you feel your child is stupid. This may sound silly, but sometimes loving, caring parents unintentionally give their kids the most negative messages; for example, "Even your little sister Stephanie can do that," "Hurry up, can't you see that the answer is ten?", "Don't worry, math was hard for me too—and besides, you'll never use it!", or "How come you got a B in math when you could get A's in everything else?"

- Find positive ways to support your child's teacher and school. Join the parent group. Offer to help find materials or role models. Accompany field trips. Avoid making negative comments about the teacher or the school in front of your child—your child needs to maintain a good feeling about the school.

- Try not to drill your child on math content or create hostilities by insisting that math work be done at any one specific time or in a specific way. Don't use math work as a punishment. Parents and adolescents have enough things that may create friction without adding math to the list.

- Model persistence and pleasure with mathematics. Include enrichment, recreational math in your family routine. Try to introduce math ideas (with a light touch!) at the dinner table, or while traveling, even to the grocery store.

Helpful Hints reprinted by permission of Lawrence Hall of Science from Family Math *by J. K. Stenmark, V. Thompson, and R. Cossey.*

17 Fun with Science in Your Community

Phyllis Katz
The Hands-On-Science Program
Rockville, Maryland

"Making time for . . ." has become an important part of modern life. Today, we schedule time for exercise, time to be with our children, or a few moments of peaceful solitude. Not so long ago, we didn't have to "make time" to tone our muscles with special equipment and gym memberships. Normal activities were inherently physical. Walking, planting, gathering, chopping wood, and building and repairing a place to live (or moving to another) not only kept our bodies in shape, but our senses sharp. We performed daily experiments upon which our survival depended. We tried dragging or pushing. We tried raw and cooked. We tried stones and clay and metal. Children tagged along, practicing skills as we made our way through history, until now. We live in an environment of such complex specializations that we must "make time" to use our bodies to keep them healthy, word process to communicate, and food process and microwave to cook. We "make time" to give our children attention because often they can't tag along. We send our children to school to teach them how to solve problems. In many ways, our activities lack survival value. How much does the child know about seeds, weather, and soil when he or she opens a package of frozen vegetables? As we make time for exercising our bodies, we need to make time for exercising our minds.

The Need for Problem Solving

Problem solving is course work for our children. Say "experiment" and visions of test tubes, data sheets, and laboratory settings come to mind. School is work for our children. They see it that way. They are given goals, progress reports, and systems of accountability by which they are tested and compared to themselves, their neighbors, and their national counterparts. Children divide school and play as adults divide work and play—sharply. Without question, they need discipline and responsibility, but unfortunately, children equate pleasure with play, not pleasure with work.

The Importance of Play

"Let's play," the kids shout. We've heard it so many times that often we don't think about its meaning. What are we saying when we invite play? Maybe it's the challenge of strategy on a chessboard. It could be the physical competition of baseball, hockey, or volleyball. It might be building a sand castle, or painting, or coloring with crayons on paper. Perhaps it's an invitation to share a pretend world where one sets the stage, picks the characters, and controls all the action. Whatever the choice, play involves some planning, coordination, trial and error, materials manipulation, peer discussion, evaluation, and retrial. Play is self-initiated. It is fun because it's voluntary and terminated when the pleasure principle isn't working. It is an early and lasting form of informal science education, if you choose to look at it that way. Things happen differently on blacktop than on concrete. Some balls are prized for their bounce. Our formal education is a few short, intensive years but informal education goes on for the rest of our lives, in our hobbies, sports, and work.

If we combine our need to "make time" for play with time for informal science education, we need to plan and find a place to make it happen. The Hands-On-Science Afterschool Enrichment Program was born from these needs.

The Hands-On-Science Outreach Program is a choice for the nonschool hours of children and their families. We often say that Hands-On-Science classes are to school science what piano lessons are to the music curriculum—an opportunity to do more for those who want it. Local agencies and individual PTAs find financial support for those families who need it. We have a cycle of activities so children can, if they want to, participate from the time they are four years old until they finish sixth grade, without repeating. Or they can, and do, float in and out of the series when particular topics stimulate their interest.

The Beginnings of Hands-On Science

Montgomery County, Maryland, while perhaps not unique, is unusual. Just north of the nation's capital, its well-educated population demands and supports a large, high-quality, countywide school system of more than 100 elementary schools. In addition to the individual Parent-Teachers Associations (PTA) that support each school, the PTA operates on a county level, where it powerfully lobbies the county for education tax dollars and communicates with the Board of Education and the school administration. In the 1970s, when budget cuts forced foreign language out of elementary schools, the Montgomery County Council of PTAs created Educational Programs, Inc., a nonprofit educational corporation, to pay a director and teachers to offer foreign language before or after school as a fee-based option. When "recreational science" was introduced, therefore, a mechanism was in place to accommodate it.

The Hands-On-Science (HOS) afterschool enrichment classes run in three (fall, winter, and spring) eight-week sessions each year. There are four age/grade levels (Pre–K, K–1, 2–3, 4–6) that meet once a week for an hour and are limited to 10 or 11 children.

For more than three years, the Hands-On-Science Program grew rapidly, mostly by word-of-mouth, parent-to-parent. The metropolitan Washington area is somewhat transient. People move in and out according to the political climate, government contracts, or military assignments. Before long, people were asking to take HOS with them to their new locations. In 1984, Hands-On-Science Outreach, Inc. (HOSO), was formed as another nonprofit company to provide HOS classes outside of Montgomery County. In 1985, the National Science Foundation awarded a grant to pilot the HOS Program in selected communities across the country.

Mixing Play with Problem Solving

The Hands-On-Science Program makes time for play. Playing pieces or tools are selected and organized by themes to give children experience with the phenomena or the processes we call science. The themes help children group their experiences, see the trends, and draw conclusions to form experiential theories. The titles are playful and enticing. "The Toymaker" is a series on simple machines for second- and third-graders. Children make simple toys to understand the function of inclined planes, wedges, wheels, and axles. The children get materials and guidance but they aren't graded on the quality or frequency of their responses. They are free to "play" with the materials—to see what happens if they add cold water instead of warm water to a yeast mixture, or to watch how the movement of a paper clip affects the stability of a paper airplane flight. They are encouraged to cooperate, not necessarily to distinguish themselves as individuals. All program materials go home with children so that experiments and activities can be repeated or explained to family or friends.

The HOS activities are chosen for their age-appropriate skill and content level, as well as their safety and simplicity. To keep HOS classes within the reach of most children, the costs are kept low, and this means shoestring science. We use lots of paper plates, cups, glue, plastic mirrors and thermometers, disposable measuring cups, rock samples, flip disks, whirligigs, and kite string and shoestring. While some push-button, computer-oriented children have been disappointed at first, they usually become actively curious when they are challenged to do such things as mix a proportion of glue, glitter, and water so it will flow to make designs on a paper cone pendulum.

The acceptance and positive use of what doesn't work is another important lesson in HOS. Failure, in its experimental sense, is the elimination of one pathway to a goal. It is a positive action. "Good, we know this won't work, so we have to try another way." A little frustration can go a long way to encourage ingenuity. HOS teacher training emphasizes positive attitudes and good self-images. The HOS experiments and activities can be familiar activities approached from new angles. They can resemble arts and crafts, music, game playing, or storytelling. We look for experiences to which science can be applied. We seek to keep that sense of wonder in children that is so often eliminated in their quest to please adults. The children explore rather than seek to memorize in after-school, informal science.

The Teachers

Hands-On-Science teachers are mentors more than they are formal teachers. They are not experts, but leaders who present situations and maintain enough order for safety and a good use of time. The typical HOS teacher is a former elementary school teacher, but HOS teachers can be retired people, graduate students, freelancers, artists, gymnasts, engineers, writers—even former or present scientists. They have in common a sense of fun, energy, genuine interest, and pleasure in the way children grow and think. While some very energetic fulltime teachers work with the HOS Program, it is a general policy that they don't. Regular classroom teachers present a conflict: They work all day teaching already; they dress more formally than we like HOS teachers to dress; and most importantly, the children in their schools see them as formal educators. Most children would rather not participate than appear unknowledgeable to the teachers who grade them. So the HOS teacher is drawn from the greater community, a parent or other adult who isn't a full-time teacher.

HOS teachers attend a training program that has been refined over the years to meet the specific needs of afterschool, informal science. They learn to stimulate tired children, use behavior management techniques that keep the children productively busy but unregimented, answer questions with questions, and insist that the children exercise their brains. They learn to feel comfortable saying "I don't know." This is a very important part of the role modeling that an HOS teacher does. The courage to say "I don't know," followed by "Let's find out," or "How could we go about finding out?" is crucial to discovery. What is there to discover if you know it all, or think you should?

Because HOS classes are midday and employment is at most a few hours a week, most HOS teachers are women. This means that children see women model an interest in science. Women bring the children their science materials. Women help them set up their experiments. And women have fun doing these things. Since, for so many years, the image of a scientist has been a man in a laboratory coat, this association of women with science is welcome.

The presence of the HOS Program raises the science consciousness of a community. From the parent volunteer who coordinates the information and registration at her school, to the people who teach, and to the school administration and faculty who encourage the use of their facilities, the HOS Program involves a number of adult participants as well as children.

Kit Materials

Before undertaking the massive project of kit production, we had teachers purchase their own supplies, but it was time-consuming and inconvenient and therefore didn't always get done. For a while, we stuffed material into donated grocery and liquor boxes. These were somewhat out of place in schools where alcohol consumption is forbidden! As the program demand grew, we needed greater reliability and cost-effectiveness. At one point, we standardized the boxes and fed children pizza dinners if they helped pack the kit. Today, the materials are packaged professionally in neat, uniform boxes by a sheltered workshop.

Inside is a list of what the HOS teacher needs each week. The materials are fun, safe, and for the most part, easily found, so children can repeat the activities that interest them. Some teachers bring the entire box each week. Some store them at home or at their teaching sites and pull only what the week's activities require. It's their choice.

We have been asked to supply the curriculum, the kits, or the training separately, but we don't. They are fully integrated and one doesn't work without the others. Together, they work well.

The Place

Hands-On-Science classes, sponsored by the parents in a school, usually take place at a child's home school. Public or private, the school building is already a familiar gathering place for the children, parents' groups, and community leaders. Many jurisdictions have established "Community School" programs for better use of the buildings. They have policies and personnel to deal easily with nonschool system users. A school building, during nonschool hours, is a good place for Hands-On-Science activities. But libraries, churches, synagogues, YMCAs and YWCAs, museums, and nature centers have also been host sites. They too are interested in providing safe locations and good activities for children. Depending on the geography, demography, and politics of the community, any of these places will work.

Family Science Festival

Once HOS classes are in full swing and their presence stimulates interest, people want to find more ways to do informal science. We have developed summer programs and a Family Science Festival to meet these demands. The summer programs are offered in two-week sessions for mornings, afternoons, or both. They present more extensive explorations than are possible in an hour a week and use more space and somewhat different equipment. The Family Science Festival is an event.

We stage our Family Science Festival once a year on a winter Sunday afternoon so we can include families who have transportation or scheduling problems during the week. It is a simple happening, where we find a big, open space—the gymnasium of a community center, in our case—and group tables and lay out materials and self-guided directions. We charge an entry fee for the children. To discourage dropouts and encourage family togetherness for these science activities, parents get in free. At one end, we set up a flight runway for making and testing paper airplanes. We have a series of brain teasers in pencil boxes. There are physical activities, simple chemistry tests, and all sorts of other things to do. We use a separate room for younger children who experiment with magnets, do dinosaur shape rubbings, and roll objects down different inclined planes. Each year we add different activities to the festival for variety for returning participants. We have published a manual, *Putting Together a Family Science Festival*, for others to duplicate this type of event. (The manual is available from NSTA for $8. Write to NSTA Special Publications, 1742 Connecticut Ave. NW, Washington, DC 20009.) We'd like to see it used, perhaps, at local schools as a fundraising activity in place of, or in addition to, say, a spaghetti dinner. It's as much fun as a school carnival, requires less out-of-pocket cash for an organization, and can demonstrate the pleasure in continued, informal science.

Parents and Schools—A Partnership

All in all, the Hands-On-Science Program makes more time for science education. It brings parents and school systems closer together as a partnership in education. We have been told that children who take HOS classes ask more and better questions in their formal science classes, and teachers enjoy their increased interest. As parents become more involved through HOS, they feel more comfortable in the school and often volunteer to help the regular classroom teacher and support the inschool science program.

Hands-On-Science afterschool classes "make time" for children to find workable solutions without pressure and with pleasure. We each balance our lives differently between work and play—those things we must do and those things we choose to do. The more the former overlaps with the latter, the more harmonious our lives, the happier we are. The HOS Program aims at the pleasure of problem solving, the joys of discovery, the positive self-image and confidence derived from a thoughtful approach to trial and error. Here, in small groups, and with adults who share and model science for fun, children (and those adults) set time aside to play with science. Perhaps they will become tomorrow's scientists. Perhaps more women and minorities will feel more comfortable with pursuing science careers. Maybe children will choose scientific jobs, even if they can't be the geniuses of their professions. Perhaps HOS students will just be more inventive, flexible individuals because they have been exposed to more adults who can say shamelessly, "I don't know. Let's find out," or "I hadn't thought of doing it that way. Let's show it to the others."

Helpful Hints

- There is strength in numbers. Form a coalition of schools and parents' organizations to work together. Your community may already have useful networks. Use them.

- Check out the possibilities for facilities. Some communities are flexible and others are rigid. Some charge fees, and some don't.

- Check out insurance coverage. Schools carry liability insurance, but school users should also carry insurance. This is an expensive, but necessary, part of any program and should not be overlooked.

- Make long-range plans. Once you have sparked the interest in your community, you will want to be able to feed it. What are your community's resources?

- Prepare a realistic budget. Will you use volunteers, older students, or teachers for your leaders? Is there a local pay scale established by the community schools or unions? What will it take to prepare and distribute the class materials?

- Consider safety carefully in the preparation of your activities. Not all informal science activities found in the best books are appropriate to groups of children exploring with nonspecialists.

- Become aware of existing resources in your community. Stay in touch with your local colleges and universities. Even if they are not currently administering an informal science program, they are in the pipeline to receive information on successful programs. Check with museums. They are, by their nature, institutions of informal learning. Some already do and many could provide the training necessary to do neighborhood informal science education training on a regular basis.

- Hands-On-Science Outreach classes are operating in 21 sites in the United States. For more information, contact the Hands-On-Science Program, 4910 Macon Rd., Rockville, MD 20852; (301) 881-1142.

18 State Academies of Science: A Partnership with Programs for the Gifted

Lynn W. Glass
Iowa Junior Academy of Science
Iowa State University
Ames, Iowa

> Intellectual activity anywhere is the same, whether at the frontier of knowledge or in the third-grade classroom. What a scientist does at his desk or in his laboratory, what a literary critic does in reading a poem, are of the same order as what anybody else does when he is engaged in like activities—if he is to achieve understanding. The difference is in degree, not in kind. The schoolboy learning physics *is* a physicist, and it is easier for him to learn physics behaving like a physicist than doing something else (Bruner, 1963).

This quote from Jerome S. Bruner characterizes more than a half-century of activity by the National Association of Academies of Science (NAAS). In a letter to the membership in 1928, Howard Enders, secretary of the Academy Conference, the forerunner of the NAAS, said, "The future of our scientific culture depends largely on finding promising young people and helping them. Is it possible for affiliated academies to take up this question, and how may the Academy Conference be useful in this connection?" By 1929, the junior academy of science movement was becoming the major thrust of many state academies of science. By 1930, state academies in Alabama, Illinois, Indiana, Iowa, Kansas, Pennsylvania, Texas, and West Virginia had adopted the junior academy movement. From these early efforts, state academies of science have become essential partners of school programs for gifted science students.

The primary purpose of the early junior academy movement was to create and to cultivate an interest in scientific work among high school students. These early founders recognized that many students would not have an opportunity to go to college and that the junior academy should, therefore, teach students the scientific ideal of seeking truth (Bilsing, 1934). In addition, the junior academies were a way to discover and to develop students who had the potential to become future leaders in science. Early handbooks recommended close cooperation between state academy scientists and junior academy members to help junior members conduct and fund their research. The handbooks encouraged cooperation further by recommending that junior academy programs be an important part of the annual meeting of their state academy of science.

Junior Academies Today

Today, NAAS promotes the common aims of the various state academies and the American Association for the Advancement of Science (AAAS). The NAAS represents over 90 percent of the more than 50 science academies in the United States and has a combined membership of nearly 35,000. Since 1962, one of the major activities of this organization has been supporting and sponsoring the American Junior Academy of Science (AJAS), the modern-day descendant of the early junior academy efforts.

Although present-day junior academies are organized diversely, a unity of purpose exists—to provide enriching scientific experiences for all youth, to identify students who are academically talented in science, to develop scientific talent, and to allow students to present results of their scientific endeavors in an atmosphere that closely resembles what is available to senior scientists (Baker, 1970). These purposes have remained relatively unchanged for more than half a century.

In perhaps the pinnacle of junior academy activities, high school students present research papers at the AJAS annual meeting (held concurrently with the AAAS annual meeting). Junior academy youth from throughout the nation share their scientific talent by displaying poster papers and presenting oral scientific reports. In recent years, the male-to-female ratio of students presenting papers has been 5 to 4, a fact indicating strong participation by scientifically gifted young women. Paper titles at a recent meeting include Isomerization of Phosphoenolpyruvate carboxylase (PEPC), The Study of the Phagocytosis of Ureaplasma urealyticum by Neutrophilic White Blood Cells, Microbial Decomposition of Organic Residue in Soils, C-V Measurements of Gallium Oxide and Silicon Dioxide on Gallium Arsenide Wafers, and A Study of the Effects of Noise Upon the Concentration of Adolescents. The titles, as well as the presentations, reflect serious thought and effort put into these research efforts. No formal judging of posters or papers takes place at this meeting.

Participation in the AJAS/AAAS annual meeting is not all work. Youth are encouraged to attend scientific presentations by AAAS members. Each year during this meeting, the AAAS conducts a one-day Youth Symposium in which junior academy youth and leading scientists participate in large and small group sessions. Besides scientific activities, each meeting features tours of area scientific laboratories, museums, or other points of local interest. Time to share experiences is one of the most important aspects. Meeting people at the Youth Symposium is a valuable part of the maturing process for these young scientists.

At the grass roots level, state and local science academies foster scientific research by precollege youth. Most state academies have created regional districts to enhance service and to create a greater identity with academy members. District directors are usually high school teachers who are members of the state science academy and who have successfully initiated and supported research done by high school students.

A broad program of activities provides enriching scientific experiences for all junior academy members. Teachers, science club leaders, and junior academy directors take leadership training sessions. "How To" workshops for youth include selecting a research project, finding a suitable place to conduct research, locating helpful research scientists, and organizing and presenting research results. Many state academies maintain a roster of capable and interested scientists who give their time freely to encourage and help youth interested in exploring scientific phenomena.

Good communication among school personnel, scientists, and students is essential to establish and maintain excellent programs for gifted science students. State academy newsletters are forums for sharing ideas for research projects, announcing meeting dates, and communicating with junior and senior academy members with similar interests. Many state junior academies publish abstracts of completed student research projects, an important outlet for highly creative talents. The need for self-motivated continuing education—informal education—is made clear to students who watch the process through the junior academies.

Most junior academies have an annual meeting, usually in conjunction with the senior academy of science, to present student research and to conduct academy business. Senior scientists perform an important service by evaluating presented research. Usually, outstanding students from the state meeting are invited to attend the American Junior Academy of Science meeting, often with full financial support from the sponsoring state academy.

Most state academies have sources of money for encouraging and supporting student research. Since 1961 the AAAS has operated a program of competitive grants (AAAS Student Research Grants) to stimulate and improve science research by high school students. This program is administered through state junior academies. Active junior academies enjoy support from a variety of sources, including electric and gas utilities, communication firms, manufacturers, high-technology firms, foundations, civic and service clubs, and individuals. These funds usually are distributed on a competitive basis as either "startup" grants for beginning students or "continuation" grants for advanced students. The awards range from $25 to $200 and can be used for expendable materials or equipment that schools do not normally have.

Each state gives students awards for completing research projects successfully. Some states divide awards into categories by grade level or subject. Awards are usually plaques, trophies, medallions, certificates, university scholarships, or trips to scientific meetings, including the AJAS annual meeting or the International Science and Engineering Fair. Banquets presided over by a governor or scientist are a popular form of recognition in some states. The AAAS recognizes student research achievement by giving honorary high school student memberships; some state academies also provide honorary memberships.

Without the hard work of countless teachers, many of these research programs would not succeed. Many state academies have begun to recognize and honor teachers in the same way that they recognize students: with scholarships, trips to national meetings, recognition banquets, and certificates.

Renzulli's Enrichment Triad Model

What the junior academies of science try to accomplish is consistent with most models for working with talented students, who are above the norm in intellectual ability, specific academic aptitude, creative or productive thinking, leadership skills, visual and performing arts, or psychomotor skills. Researchers estimate that three to five percent of all school-age children fit this definition (Thomas, 1976).

Because these students have high capabilities, educational programs need to be fluid and dynamic. Teachers must serve as facilitators rather than as directors of learning by providing new learning opportunities that challenge the students. Joseph Renzulli developed the Enrichment Triad Model, a theoretical model for talented and gifted students that is the basis for junior academy programs and many school-based programs.

The rationale behind the Enrichment Triad Model is Renzulli's characterization of giftedness and the ways eminent persons create products useful to society. He believes that creativity and the effort students put into a project, not just high IQs, should be the criteria for admitting students into special enrichment programs. Advanced course work may not be sufficient or appropriate if it is intended for all children. For example, a student could be advanced two grade levels because he or she is gifted. The course work for this new grade placement may not be appropriate because it has not been designed specifically for gifted youth.

According to Renzulli, with some advanced course work everyone marches to the beat of the same drummer, albeit at different rates. Furthermore, Renzulli believes learning should result from actual inquiry, and students need to work in the same ways as professionals in specific fields (1977). The only difference between a student's and a professional's work *is* in degree, not in kind (Bruner, 1963).

The Enrichment Triad Model consists of three activity types that provide a unique opportunity for schools and junior academies to work together to educate the talented and gifted. Type I activities involve students with people, interests, and avenues outside the normal curricula. These activities include listening to guest speakers; taking field trips; participating in special assemblies; or watching films, videotapes, or television programs. Very little structure exists in these activities; they are designed to give students a bona fide idea of what an independent research project might include. Students explore activities purposefully, and after a given time period each student analyzes his or her own experiences and comes up with alternative suggestions for further study. Elementary and secondary teachers arrange these activities, often in cooperation with local or state academies of science, to help students become aware of the scientific world. Teachers can use type I activities for all children, such as to begin a new science unit or to suggest science fair or club projects.

Type II activities develop high-level thinking processes. Materials, methods, and instructional techniques used in these activities improve the student's creativity and problem-solving, decision-making, or thinking skills. During this phase, talented and gifted students meet with the resource teacher at least once a week to learn how to use the computer, gather and sort local rocks and minerals, conduct an interview, organize and interpret data, or design an experiment. Benjamin Bloom's *Taxonomy of Educational Objectives* (1956) and J. Paul Guilford's Structure of the Intellect Model in *The Nature of Human Intelligence* (1967) are helpful aids for preparing these activities. Workshops sponsored by local or state science academies can help teachers focus on skills needed to conduct scientific investigations. Some teachers use science academy resource person lists to help them with type II enrichment activities.

In type III activities, which are a marriage of type I and type II activities, each talented and gifted student conducts an individual interest project. The student actually investigates a real problem by using appropriate methods of inquiry. Examples of type III activities include topics (on areas such as human medicine and behavior, astronomy, physics, biology, and chemistry) presented at AJAS meetings. The examples I listed previously show that type III students become experts in their field and develop creative products worthy of sharing with interested audiences. When students function on the cutting edge of science, teachers can lighten the load by helping to identify resources, but cannot direct learning. At this stage, teachers must seek help outside the classroom; most state academies are capable and willing to provide this kind of help. Many successful, school-based talented and gifted programs rely upon partnerships with state science academies, businesses, and industries to accomplish their objectives. Through these shared partnerships, schools can receive advice from academic and industrial scientists, use laboratory facilities, or borrow expensive or unusual equipment and materials (Glass, 1983).

Most students, when asked what they did with their completed research project, say, "I put it in my folder and waited for the teacher to grade it." But as a final activity in the Enrichment Triad Model, students demonstrate their newly acquired professional skills by communicating the results of their research projects. Students learn to inform, entertain, or influence a relatively specific audience, just as composers write symphonies for lovers of classical music; consumer advocates carry out research to bring about legislation; and scientists communicate their findings to others. In Renzulli's words,

> It seems nothing short of criminal negligence to do everything in our power to encourage youngsters to develop the highest levels of creative thinking, and then overlook the step which is a natural consequence of real-world productivity—communication of results to appropriate audiences (1977).

It is our responsibility as teachers to find real audiences for students to communicate their findings. Local and state junior science academies, in concert with the AJAS, provide a real-world forum for students to develop the highest levels of creative thinking and to share their research results.

References

Baker, C.L. (1970). *Academy conference 1926–1970.* Columbus, OH: National Association of Academies of Science.

Bilsing, S.W. (1934). Science clubs in relation to state academies of science. *Science Education, 18,* 162–167.

Bloom, B.S. (1956). *Taxonomy of educational objectives: Handbook I cognitive domain.* New York: David McKay.

Bruner, J.S. (1963). *The process of education.* New York: Vintage Books.

Glass, L.W. (1983). Business and industrial support of high school science education. *School Science and Mathematics, 83,* 91–95.

Guilford, J.P. (1967). *The nature of human intelligence.* New York: McGraw-Hill.

Renzulli, J.S. (1977). *The enrichment triad model: A guide for developing defensible programs for the gifted and talented.* Mansfield Center, CT: Creative Learning Press.

Thomas, D. (1976). Gifted and talented children: The neglected minority. *NASSP Bulletin, 60,* 21–24.

Helpful Hints

- Contact your local junior academy of science director. Request the name of your director from Lynn E. Elfner, Archivist, National Association of Academies of Science, 445 King Ave., Columbus, OH 43201, or R. Dean Decker, Director, American Junior Academy of Science, Biology Department, University of Richmond, Richmond, VA 23173.

- Request a copy of your local junior academy handbook. These handbooks contain rules for local junior academy activities, guidelines for research grant applications, and dates for student and teacher workshops and meetings.

- Request lists of active students and teachers near your school. Active students and teachers are contagious. Their enthusiasm spreads to others.

- Participate in meetings and workshops run by your local science academy or junior academy. If you do not have an active local or state junior academy of science, take the initiative and start one. After all, active junior academies exist because teachers like yourself saw the need for them and invested the energy necessary to create an organization that helps others.

- Encourage students to develop scientific talent by collaborating with scientists in your geographical region to initiate, conduct, and report independent scientific research.

19

Toys, Books, and Other Science Gifts

E. Wendy Saul
University of Maryland
Catonsville, Maryland

Mrs. B.'s fourth-grade class had spent the last several weeks working on a model of the "village of the future." The teacher was impressed with the project and wanted to contribute to her students' efforts. She had an idea for a "gift"—batteries, wire, small light bulbs, and directions for completing a circuit. The children wired the houses, the hospital, and the airport, and were apparently as thrilled by the gift of light as any resident who first saw the crews of the Tennessee Valley Authority making their way down the road.

Every year countless newspaper articles and television commentators bemoan the sorry presents we provide for future citizens. In few instances, however, do they suggest alternatives to purchasing the latest gizmos—weaponry for laser tag or dolls who repeat kindly messages directed at no one.

Teachers, parents, librarians, scout leaders, and museum personnel all search for gifts that not only please the child, but stir the imagination. Excellent science gifts are available, but simply buying the gift does not ensure that it will fulfill its purpose. One must find the right gift for the right child at the right time, and then introduce it with care, curiosity, and concern. What an adult says about or does with a gift is as important as the gift itself.

What is a "science toy?" I use the term to mean any item that helps in the playful (though perhaps systematic) observation and classification of the natural world. Science toys are more than chemistry sets and toy microscopes. I include bird feeders, balances, protractors, magnifiers, building sets, flower presses, and, of course, books.

Matching the Gift to the Child

Science toys and books are not just for children already interested in science. The good science shopper finds items that help children see the connection between their interests and scientific knowledge and processes. For Judy, who likes art and beautiful things, a perfume-making kit may be just the thing. If Judy especially likes the odors extracted from lavender or chamomile, she might even plant a small garden. From there, the careful gift-giver might move on to a book about wildflowers and herbs. Working with greenhouse plants is different from working with outdoor plants. Here again, tools and books become necessary items for engaging in science play.

Aaron, on the other hand, may not show such keen enthusiasm for plants, or on the surface, for anything else. Though his classwork comes easily, Aaron is unhappy—he has nothing "interesting" to do with his free time. But when his school principal invites members of the American Radio Relay League (ham radio operators) to speak at a school assembly, Aaron is fascinated. He writes for more information and becomes an active member of a group that appreciates and uses his abilities.

The point here is that finding the right item for the right child at the right time involves more than reading a box that claims "Fun for All" or "Recommended for Ages 10 and above." One must know the children and their individual interests; one also must understand, to some extent, the skills and dispositions that various sciences, or science items, require. For instance, wildlife watching requires patience; the observer has little control over what happens. In addition, finding good places to sit and watch, and coming up with alternative explanations of why a certain insect seems to be attracted to particular flowers, or theorizing where it will alight next takes a speculative creativity. An electronics set, in contrast, requires a different kind of attention. This gift is best for the child who enjoys following directions. Mastery is evident—either the constructed item works or it doesn't. Feedback is immediate and satisfying.

Don't Shoot for the Stars—Yet

In working with science toys and books, one point is clear. Although there are some "bad" materials—they don't work or are dangerous—worth is largely determined by what the child brings to the experience. Suppose Jamie, who has shown no previous interest in astronomy, requests a telescope. If Jamie really wants to study astronomy, however, a book on naked-eye astronomy, a star chart, and a pair of wide-angle binoculars will serve her purpose better. Not only can binoculars produce excellent views of the moon, but they can also be used for birdwatching or a trip to the ballet. People who buy telescopes are often frustrated to find that although the moon and some planets make good targets, stars through a telescope usually look like stars without a telescope, only brighter. The fascination of astronomy comes in finding one's way around the heavens, observing changes over a period of time, and relating what one knows about an object to that speck in the night sky.

The rule of thumb is not to inflate a child's expectations unrealistically. Chemistry sets, for instance, no longer allow the user to make mini-explosions or stick bombs. (The old Gilbert Chemistry Cupboard of your youth is no longer available!) On the other hand, some sets challenge the child to understand that chemistry involves varying conditions in order to produce specific changes. Some children find pleasure in seeing that changes are not magic, but predictable and replicable.

The manufacturers of science products are at least partially to blame for the "failure" of science play. Store-bought microscopes, like telescopes, build on the misconception that increased power means increased optical clarity. Unsuspecting consumers set up their 1200X device, for which they have paid twice the price of a comparable 300X microscope, and are disappointed to find an image blown up beyond recognition and so dark that no detail is evident. Only the prepared slides from the manufacturer can be viewed reliably. Children eager to study salt crystals, dirty fingernails, pond water, and dandruff are told, in effect, that their interests aren't the stuff of which science is made. Young children do much better with a hand magnifier or an illuminated pocketscope.

Books, on the other hand, are constantly reviewed and evaluated, and thus standards appear to be less of a problem. Except for the legal aspects of safety, little is made of a given toy's poor instructions or shoddily constructed parts. The recently published *Science Fare: An Illustrated Guide & Catalog of Toys, Books and Activities for Kids* (Saul, 1986) is one of only a few publications (along with *Consumer Reports*) to take toy reviewing seriously.

Choosing a Book

I count books among the list of science toys for three reasons. First, a number of books, such as Jim Arnosky's extraordinary *Secrets of a Wildlife Watcher* (1983), or Grace Mamor Spruch's *Such Agreeable Friends: Life with a Remarkable Group of Urban Squirrels* (1984) are every bit as helpful in observing as a pair of the best binoculars. Field guides are also indispensable aids for the would-be naturalist.

Second, books often provide directions for creating toys. For example, Mollie Rights' lively text *Beastly Neighbors: All About Wild Things in the City* (1981) gives directions for constructing a better terrarium than those commercially available. Similarly, Vicki Cobbs' *Chemically Active* (1986) functions as an excellent chemistry set for those who engage in the suggested experiments with readily available materials.

And third, books provide the reader with a model of scientific curiosity and an approach to thinking about the physical world perhaps not available at school or at home. Biographies such as *Six Little Chickadees: A Scientist and Her Work with Birds* by Ada Graham (1982) describes the motivation and activities of Cordelia Stanwood in a way young people can comprehend, while a classic work such as Michael Faraday's *The Chemical History of a Candle* (1960), first published in 1861, is a model of scientific reasoning and skill.

Books on nature, physical science, and mathematics are so varied and numerous it is difficult to choose the best examples. A work such as Virginia Lee Burton's *Life Story: The Story of Life on Earth From Its Beginning Up to Now* (1962), though not entirely up-to-date in scientific detail, is so rich in its sense of geologic time and such a good invitation to science that it is a welcome addition to any library. Other presentations use children's own queries as the organizational format of the text; see, for instance, Roy Gallant's *101 Questions About the Universe* (1984), which is based largely on answers to actual questions children ask at the planetarium where he works.

In the past, children were given almost any science book that was accurate and accessible. Books were written for those who already expressed an interest in (or were assigned) a topic. Recent works, though, have consciously sought a hook to intrigue children and awaken their scientific interests. Have you ever wondered, for instance, what goes on beneath a busy street corner in a large metropolitan area? Read David

Macaulay's *Underground* (1976) for an indepth, visual description of what takes place there. Or try Wyatt Blassingame's *The Little Killers: Fleas, Lice, Mosquitoes* (1975) for amazing tales of the major historical events wrought by these tiny parasites. Moreover, the illustrations and design features of children's science books compare with the most stunning picture books available. Check out, for instance, *Your Future in Space: The U.S. Space Camp Training Program* by Flip and Debra Schulke and Penelope and Raymond McPhee (Schulke et al., 1986) or Patricia Lauber's *Volcano: The Eruption and Healing of Mount St. Helens* (1986). Also worth mentioning is Seymour Simon's spectacular new series on the solar system (1985).

Science books, which are sold predominantly to libraries, are reviewed by a press accountable to both the public and the scientific community. Sources of such reviews are given at the end of this chapter.

Curiosity: The Greatest Gift

What can teachers and parents do to promote aggressive curiosity? I am not talking about an anxiousness to touch, although that may well be involved, but rather an interest in grasping the principles underlying the organization of scientific fact. Although some children seem to be blessed with scientific aptitude the way others demonstrate artistic talent, there are things adults can do to foster whatever gifts are in evidence.

Nothing is more important than encouraging children's questions. Unfortunately, the same adults who are delighted to answer their children's queries about definitions or spellings of words often inadvertently discourage science questions. A child who asks why a crusty coating of snow is on the ground on winter mornings deserves more than a story about Jack Frost. If you don't know the answer to a question, remember you are not alone—even the most scientifically literate individuals are not familiar with the entire body of knowledge we call science. Visit your local library, ask experts, undertake experiments with your children; but most importantly, show them that their queries are worthy of adult attention.

You can also model questioning. In listening to the questions of scientists, certain recurrent queries can be noted: *why, how, how much,* and *what would happen if* are the scientist's stock-in-trade. Make them yours, too.

For instance, scientists assume that things in the physical world have a *function.* If you are looking at a llama, you might ask, "Why is it woolly?" or "What are the functions of its hooves?" *Why* questions don't have to be limited to natural objects; while studying a faucet, for instance, invite children to consider what function each part serves. You can also ask questions about how things work. How does the woolly covering of an animal provide warmth? What happens if this seemingly useless ring on the faucet is removed? Numerical descriptions also interest scientists. How long does the typical llama live? How long does the mama llama take "to grow a baby?" How does that compare to the gestation period of a goat? How long will it take to fill this sink with water? How long will it take to siphon the same amount of water? Scientists like to play with variables as a way of questioning how things work. In biology, this often leads to questions about genetics and adaptation. In other instances, it may lead to the construction of an experiment. Which takes longer to freeze, orange juice or water? Will baking soda fizz the way baking powder does when added to vinegar? Will lemon juice work as well as vinegar?

Assertiveness also figures into this formula. I vividly remember watching two children approach a chemistry set. One, who had already identified herself as science-interested, almost ripped into the kit. She wanted to know exactly what each of the chemicals would and would not do, and this overriding question, which she was able to verbalize, helped her decide which experiments to undertake when. The other child was less sure of himself. He saw the instruction guide as law, and any alteration or extension of it as a violation. When a particular experiment didn't work, he was disappointed, perhaps even angry, and put the set away. Although he was unsure of where the "blame" lay—was it the set's problem or had he done something wrong?—he surely had no interest in further chemical explorations. For the first child, the same "failures" were deemed "interesting."

Finally, the well-timed introduction of fact or theory often helps children formulate their own questions. For instance, if a young child is given a model of a woolly mammoth, she typically makes monster noises while skipping it along the carpet. After reading a book such as Aliki's *Wild and Woolly Mammoths* (1977), however, the play often reflects scientific knowledge. The child moves the model more "mammothlike," and the dramatic story about finding the frozen beast in Siberia may even be reenacted. As the child needs to know more to make the story her own, her questions develop.

Combining Resources

The value of pairing a toy and book can hardly be overstated. Through the union of books and toys, science, knowledge, and experience are connected. That such pairings have not been regularly suggested elsewhere is testimony to the ways toys and books, as well as hands-on science and explanation, have been truncated in our schools, in the marketplace, and finally, in our minds.

Professor Lazer Goldberg, in a recent Library of Congress symposium on "Children, Science, and Books," said, "We educate children who will live most of their lives in a time we cannot describe with any degree of accuracy. It is unlikely that what we *tell* children today will be completely useful, let alone 'true,' 40 years from now" (in press). But interest in exploring the physical world and confidence in one's ability to engage in scientific play can only serve the child well in years to come. If playthings can be used to invite science questions, the gifts you have given will be gifts for life.

How To Find Science Literature for Young People

There are several highly regarded sources for keeping up with children's books in general, some or all of which your local library will have. The following journals, though broad-based in their coverage, include reviews of children's science books:

Booklist, 22/year, $51 (American Library Association, 50 E. Huron St., Chicago, IL 60611)

Bulletin for the Center of Children's Books, 11/year, $14 (5850 Ellis Ave., Chicago, IL 60637)

Hornbook, 6/year, $32 (Park Square Bldg., 31 St. James Ave., Boston, MA 02116)

Kirkus Reviews, 24/year, $65 (200 Park Ave. S., New York, NY 10003)

School Library Journal, 11/year, $56 (R. R. Bowker, Subscription Department, P.O. Box 1978, Marion, OH 43305-1978)

The following periodicals for science teachers include a healthy smattering of book reviews:

Science and Children, 8/year, $43 (National Science Teachers Association, 1742 Connecticut Ave. NW, Washington, DC 20009)

The Science Teacher, 9/year, $43 (National Science Teachers Association, 1742 Connecticut Ave. NW, Washington, DC 20009)

Science Activities, 4/year, $35 (Heldref, Albemarle St. NW, Washington, DC 20016)

Dedicated *primarily* to reviewing science literature, the following books, periodicals, and award lists get my vote for "best bets":

Best Science Books for Children, compiled and edited by Kathryn Wolff, Joellen M. Fritsche, Elina N. Gross, and Gary T. Todd (AAAS, 1983); Children's Catalog (H. Wilson, published annually)

The Museum of Science and Industry Basic List of Children's Science Books 1973-1984, compiled by Bernice Richter and Duane Wenzel

(American Library Association, published annually)

Science Books for Children: Selections from BOOKLIST, 1976–1983, edited by Denise Murcko Wilms (American Library Association, 1985)

Appraisal, 4/year, $24 (605 Commonwealth Ave., Boston, MA 02215)

The Kobrin Letter: Books about Real People, Places and Things, 7/year, $12 (Dr. Beverly Kobrin, 732 Greer Rd., Palo Alto, CA 94303)

Science Books and Films, 5/year, $32 (AAAS, 1776 Massachusetts Ave. NW, Washington, DC 20036)

Scientific American publishes an annotated list each December of child-oriented science gift books written by Philip and Phyllis Morrison ($2.95 at newsstands)

For free lists of award-winning books, send a self-addressed, stamped envelope to:

Children's Book Guild Non-Fiction Award (c/o Washington Post, 1150 15th St. NW, Washington, DC 20017)

New York Academy of Sciences (2 East 63rd St., New York, NY 10021)

Outstanding Science Trade Books for Children (Published by Children's Book Council and National Science Teachers Association; available from Children's Book Council, 67 Irving Place, New York, NY 10003)

References

Aliki. (1977). *Wild and woolly mammoths.* New York: T.Y. Crowell.

Arnosky, J. (1983). *Secrets of a wildlife watcher.* New York: Lothrop Lee and Shepard.

Blassingame, W. (1975). *The little killers: Fleas, lice, mosquitoes.* New York: Putnam.

Blocksma, Mary, & Blocksma, Dewey. (1986). *Space-crafting: Invent your own flying spaceships.* New York: Prentice-Hall.

Burton, V.L. (1962). *Life story: The story of life on Earth from its beginning up to now.* Boston: Houghton Mifflin.

Cobb, V. (1986). *Chemically active.* New York: Lippincott.

Faraday, M. (1960). *The chemical history of a candle.* New York: Viking.

Gallant, R. (1984). *101 questions about the universe.* New York: Macmillan.

Goldberg, Lazer. (in press). E. Wendy Saul (Ed.). *Vital connections: Children, science, and books.* Washington, DC: Library of Congress.

Graham, A. (1982). *Six little chickadees: A scientist and her work with birds.* New York: Four Winds.

Lauber, Patricia. (1986). *Volcano: The eruption and healing of Mount St. Helens.* New York: Bradbury.

Macauley, D. (1976). *Underground.* Boston: Houghton Mifflin.

Rights, M. (1981). *Beastly neighbors: All about wild things in the city.* Boston: Little Brown.

Saul, E. Wendy, & Newman, Alan. (1986). *Science fare: An illustrated guide & catalog of toys, books and activities for kids.* New York: Harper & Row.

Schulke, F., et al. (1986). *Your future in space: The U.S. Space Camp Training Program.* New York: Crown.

Simon, Seymour. (1985). *Jupiter.* New York: Morrow.

Spruch, G.M. (1984). *Such agreeable friends: Life with a remarkable group of urban squirrels.* New York: Morrow.

Helpful Hints

- Invite children to bring in toys with moveable parts. Use small-group discussions to figure out, describe, and map how some items work.

- Suggest that children bring mechanical objects such as a can opener, a wire wisk, or a drill with bits. Have a child describe an object in terms of its structure and function and see if others in the class can "guess" the object described.

- Have children design exhibits for the library/media center that pair an object or experiment with a book, and invite others to experiment in similar ways. For instance, Mary and Dewey Blocksma's *Space-Crafting: Invent Your Own Flying Spaceships* (1986) could be exhibited with one of the easily constructed flying toys described. Another child might exhibit a pumice stone alongside two books—one on geology and the other on floating and sinking.

- Develop a lending library of science-related tools—microscopes, protractors, rain gauges, litmus paper, and so on. Have students design projects using these objects.

- Try to replicate some of the experiments of early scientists with the equipment on hand. Most toy telescopes, for instance, are better than the one Galileo used, and even a moderate hand magnifier of today compares favorably to that of van Leeuwenhoek.

- Encourage children to evaluate books on a given topic. After comparing various books about dinosaurs, for instance, have them talk about differences in the presentation of fact and the importance of organization and style.

- Have children generate questions BEFORE searching books for a report topic. See which questions are easily answered and which topics authors fail to address. Then encourage children to generate further questions based on their forays into the subject. Have groups trade questions, and if some can't be answered through the library, write to an expert.

- No classroom should be without field guides and an area map of the places where students have found various specimens.

Bibliography

Informal Learning

Heltne, Paul, & Marquardt, Linda A. (1988). *Science learning in the informal setting; 1988 symposium proceedings*. Chicago, IL: Chicago Academy of Sciences.

Today's problems: tomorrow's crises. (October, 1982). National Science Board Commission on Precollege Education in Mathematics, Science, and Technology.

Media

Action for Children's Television. (1984). *TV, science and kids*. Menlo Park, CA: Addison-Wesley.

Chen, Milton. (1979). *3-2-1 CONTACT, a tv series on science and technology for children 8–12: Test show evaluation*. New York, NY: Children's Television Workshop.

Friedman, Sharon, Dunwoody, Sharon, & Rogers, Carol. (1986). *Scientists and journalists*. New York, NY: Free Press.

Gastel, Barbara. (1983). *Presenting science to the public*. Philadelphia, PA: ISI Press.

Newspaper Science Sections; Testing the Mass Audience [issue]. (Autumn, 1984). *SIPI-scope*, *12*(4).

Science in the Media [issue]. (Winter, 1982–83). *SIPI-scope*, *10*(6).

Museums

Bedworth, Jacalyn. (Ed.). (1985). *Science teacher education at museums: A resource guide*. Washington, DC: Association of Science–Technology Centers.

Borun, Minda. (1980). *What's in a name? A study of the effectiveness of explanatory labels in a science museum*. Washington, DC: Association of Science–Technology Centers.

Borun, Minda et al. (1983). *Planets and pulleys: Studies of class visits to a science museum*. Washington, DC: Association of Science–Technology Centers.

Danilov, Victor J. (1982). *Science and technology centers*. Boston, MA: MIT Press.

Flexer, Barbara, & Borun, Minda. (1984). The impact of a class visit to a participatory science museum exhibit and a classroom science lesson. *Journal of Research in Science Teaching*, *21*(9), 863–873.

Grinell, Sheila. (Spring, 1988). Science centers come of age. *Issues in Science and Technology*.

McCormack, Susan. (Ed.). (in press). *Administration report & directory*. Washington, DC: Association of Science–Technology Centers.

McCormack, Susan. (Ed.). (1988). *Education report & directory*. Washington, DC: Association of Science–Technology Centers.

Oppenheimer, Frank & the Exploratorium Staff. (1986). *Working prototypes: Exhibit design at the exploratorium*. Washington, DC: Association of Science–Technology Centers.

Oppenheimer, Frank, & the Exploratorium Staff. (1987). *The Exploratorium cookbook, volumes I, II, & III.* San Francisco, CA: The Exploratorium.

Pitman-Gelles, Bonnie. (1982). *Museums, magic and children: Youth education in museums.* Washington, DC: Association of Science–Technology Centers.

Pollock, Wendy, & McCormack, Susan. (Eds.). (1988). *Exhibits report & directory.* Washington, DC: Association of Science–Technology Centers.

Tressel, George W. (1984). A museum is to touch. *1984 yearbook of science and the future.* Chicago, IL: Encyclopedia Britannica.

Activities

Black achievers in science. (1988). Chicago, IL: Museum of Science & Industry.

Children's Book Council. (published annually). *Outstanding science tradebooks for children.* New York, NY: Children's Book Council.

Cobb, Vicki. (1972). *Science experiments you can eat.* Philadelphia, PA: Lippincott Junior Books.

Flatow, Ira. (1988). *Rainbows, curved balls and other wonders of the natural world explained.* New York, NY: Wm. Morrow & Co.

Great explorations in math and science. Berkeley, CA: Lawrence Hall of Science.

Herbert, Don. (1980). *Mr. Wizard's supermarket science.* New York, NY: Random House.

Horner, John R., & Gorman, James. (1985). *Maia, A dinosaur grows up.* Bozeman, MT: Museum of the Rockies.

Reid, Theodore L. (1989). Volunteers for science. *1989 yearbook of science and the future.* Chicago, IL: Encyclopedia Britannica.

Solinini, Cheryl. (Ed.). (1987–88). *Smithsonian family learning project, science activity book.* New York, NY: Galison Books.

Zubroski, Bernard. (1979). *Bubbles.* Boston, MA: Little,Brown.

Zubroski, Bernard. (1979). *Milk carton blocks.* Boston, MA: Little,Brown.

Photo Credits

Strategies

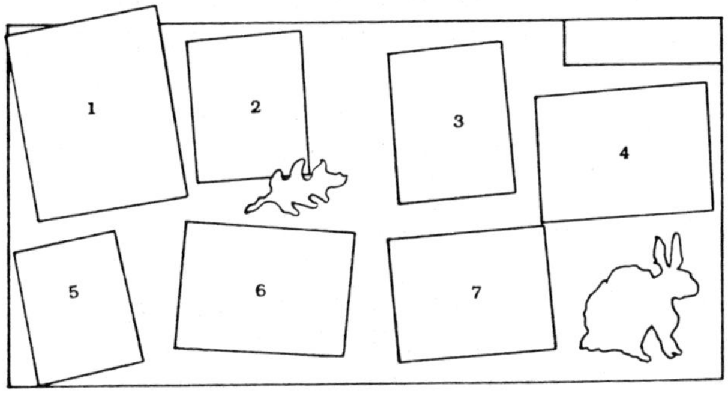

1. Children's Television Workshop
2. © J. Brown/G. Simon, Subjects & Predicates, Silver Spring, MD
3. © J. Brown/G. Simon, Subjects & Predicates, Silver Spring, MD
4. Children's Television Workshop
5. Photo File
6. The Exploratorium. Photo by Susan Schwartzenberg.
7. National Science Foundation

The Media

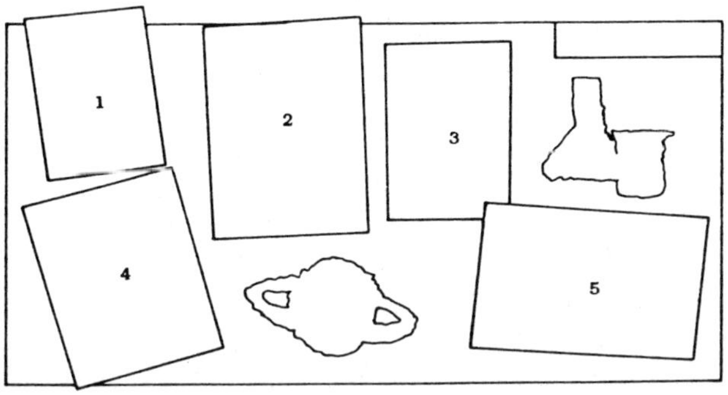

1. Children's Television Workshop
2. The Mr. Wizard Studio
3. Jerome Berkowitz
4. Photo File
5. Children's Television Workshop

Museums and Zoos

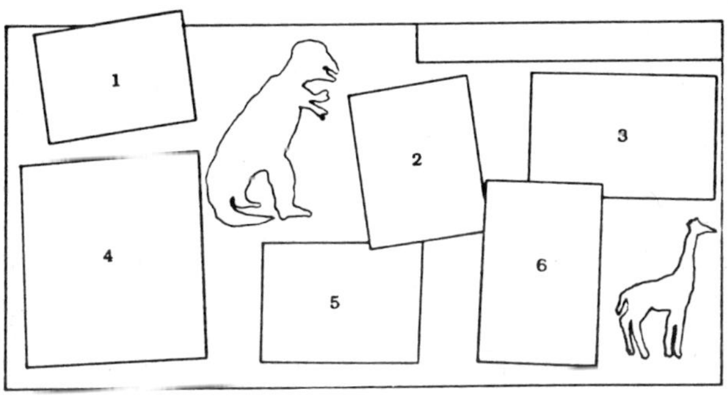

1. National Science Foundation. Photo by Esther Kutnick.
2. © New York Zoological Society Photo
3. © Zoological Society of San Diego 1987
4. © New York Zoological Society Photo
5. Woodland Park Zoo. Photo by Chris Barth.
6. National Science Foundation. Photo by Richard Howard.

Projects, Competitions, and Family Activities

1. © Joanne Meldrum
2. Karen Hollweg, Denver Audobon Society
3. Jerome Berkowitz
4. © Bruce Jennings
5. Photo File
6. Carol Luszcz, Science Talent Search